INTERACTION OF
METALS AND GASES

Distributors for The United Kingdom and Eire:

CLEAVER-HUME PRESS LTD.

(Macmillan & Co. Ltd.)

10-15 St. Martin's Street, W.C. 2

INTERACTION OF
METALS AND GASES

Vol. 1

Thermodynamics and Phase Relations

J. D. FAST

Chief Metallurgist Philips Research Laboratories
and Professor at the Technical University of Eindhoven

1965

PHILIPS TECHNICAL LIBRARY

Translated from Dutch by Mrs. M. E. Mulder-Woolcock, B.Sc., Eindhoven

This book contains x + 310 pages including 8 plates and 139 illustrations

Other editions of this Philips Technical Library book will appear in French, German
and several other languages

Printed in The Netherlands

PREFACE

Two facts have lead me to write this survey on the interaction of metals and gases: the absence of a modern book on the subject and my own long experience with metal-gas problems. The volume which lies before you bears the sub-title "Thermodynamics and Phase Relations". It will be followed by a second volume with the sub-title "Kinetics and Mechanisms". However, it was not my intention to draw a sharp dividing line between the two. Where this is helpful to the treatment I have supplemented the thermodynamic discussions in the first volume with kinetic and atomic considerations. The reverse will be the case in the second volume.

The literature on the interaction of metals and gases is very extensive. It comprises many tens of thousands of articles, so that it would require a "Handbuch" in twelve thick volumes to give a complete survey. I did not wish to attempt such a task. My aim has been to write a book which can be used as a textbook by the metallurgical student and, at the same time, may be of some use to the industrial metallurgist and chemist. Far from being "highbrow" the book handles the theoretical background as simply as possible. The theory has been illustrated by many characteristic examples which, for the greater part, are of immediate practical importance. Naturally, the treatment throughout bears the imprint of my own experience and outlook. A few items in the book are published here for the first time.

To keep the size of the book within reasonable bounds, I have been forced to omit not only many less important but also many important investigations. Naturally, my selection is open to criticism by others. However, I hope that my choice will throw some light on the whole extensive field of the interaction of metals and gases, and that the book will enable the student and the metallurgist to find their own way in this field. Each volume is intended to be complete in itself; each can be studied independently of the other.

My friend Professor J. L. Meijering (Tech. University Delft) was kind enough to read the whole manuscript of this volume before it went to press. His critical remarks have led to a number of improvements. I would like to thank not only him, but also Mrs. Mulder-Woolcock, B.Sc., for translating the Dutch text into English and Miss A. C. Leonhard for drawing the greater part of the diagrams.

Eindhoven, December 1964   J. D. Fast

CONTENTS

Chapter 5. REACTIONS BETWEEN ALLOYS AND GASES WHEREBY NEW PHASES ARE FORMED

Chapter 6. REACTION BETWEEN CARBON AND OXYGEN IN STEEL

GENERAL INTRODUCTION

Interaction with gases is of such importance to metals that it must be taken into account from their "birth" to their "death". In some respects the metal-gas problem arises even before the birth of the metal, since in many cases its manufacture involves the reduction of a metal oxide, i.e. the liberation of the metal from its combination with a gas.

Throughout every period of their "life" the commercial metals come into contact with gases. It begins in the liquid state during the production, refining and remelting. During these processes they can take up gases in solution. In particular the absorption of hydrogen due to the reaction of the hot metal with water vapour is a matter for concern. The water vapour may originate from various sources, e.g. from the walls of the crucible or furnace or from the gaseous atmosphere.

The difficulties caused by dissolved gases stem, to a great extent, from the fact that their solubility in the most frequently used metals (Fe, Al, Cu, Ni) drops sharply at the solidification point and decreases still further in the solid state. The sudden decrease in solubility on solidification can cause porosity in castings and welds. The decreasing solubility in the solid state can lead, after fairly rapid cooling of the metal, to considerable supersaturation, e.g. of hydrogen and nitrogen in steel. In some of these solid supersaturated solutions, precipitation of a metal-gas compound may occur even at ordinary temperatures. A well-known example is the gradual precipitation of iron nitride in steel containing nitrogen, a phenomenon which is accompanied by deterioration of the mechanical and magnetic properties. This gradual change in the properties is known as quench ageing, which can also be caused by the precipitation of iron carbide. In many types of steel plastic deformation is followed by another kind of ageing, called strain ageing, which depends chiefly on the interaction between nitrogen atoms and dislocations.

In supersaturated solutions of hydrogen in steel the hydrogen shows a strong tendency to precipitate in the form of H_2 molecules. This tendency can be satisfied at all points where there is sufficient space available for the H_2 molecules, provided that the temperature and the available time are such that the dissolved hydrogen atoms can reach these spots by diffusion. Precipitation of H_2 in certain lattice imperfections, which arise during plastic deformation, can lead to premature rupture of steel if the temperature and rate of deformation are favourable for this. Fracture due to H_2 precipitation in steel containing hydrogen can occur also *after* plastic deformation or under the influence of an applied stress ("delayed fracture").

Steel walls can be decarburized if they come into contact with hydrogen at high pressure and temperature. This decarburizing produced great initial difficulties in performing a number of syntheses in inorganic and organic chemistry, e.g. in the synthesis of ammonia, methanol and petrol. These difficulties have now been overcome to a large extent by the development of special types of steel.

The oxidation of metals by the air or some other oxidizing gaseous atmosphere is also a source of numerous problems. For application in the field of turbine blades, jet engines, etc., alloys have been developed which at high temperatures are relatively resistant to attack by gases (iron-chromium-nickel alloys, cobalt-chromium-nickel alloys etc.). This resistance to corrosion depends on the formation of a thin, closely adhering oxide film on the surface, which protects the metal from further attack. Selective oxidation of that component of an alloy with the greatest affinity for oxygen can, in some cases, play an important part in this respect.

Although most solid metals react with oxygen, forming an oxide phase, without an appreciable quantity of oxygen being dissolved in the metal, the commercially young metals titanium and zirconium also exhibit the ability to dissolve homogeneously large quantities of oxygen (and also nitrogen and hydrogen). Their affinity for oxygen and nitrogen is so great that without the use of high-vacuum technique it is quite impossible to produce them in malleable condition. The need for titanium in the aircraft industry and for zirconium as canning material in nuclear reactors has resulted in a high development of the technique of melting and casting these metals in a good vacuum. This development has, in turn, stimulated the vacuum melting and casting of other metals and alloys. The above-mentioned injurious action of hydrogen and nitrogen in steel, for example, can be prevented by melting and casting in vacuum.

Even at normal temperatures the possibility of interaction between metals and gases must still be taken into account. The alkaline and alkaline-earth metals oxidize in air even at normal temperatures. During pickling and plating operations atomic hydrogen penetrates into iron and steel, which may cause fracture under unfavourable conditions. Metals prepared by electrodeposition often contain large quantities of hydrogen. Many kinds of steel rust in air, enabling the iron present to return to its original state, i.e. to the state in which it occurs naturally.

The results of the interaction of metals with gases are usually undesirable. Nevertheless, there are a few cases in which this interaction can be put to a useful purpose. Examples are the synthesis of ammonia on finely divided iron by heterogeneous catalysis, case hardening of certain types of steel by the introduction of nitrogen and the use of "getters" in high-vacuum technique.

When studying the interaction between metals and gases, a distinction must be made between the final state (equilibrium state) which can be reached under the given conditions, and the rate at which this state is reached. Thermodynamics can only be of help, but it is a great help, in the study of the equilibrium states. This book deals primarily with the thermodynamic aspects of the interaction between metals and gases. The kinetic and atomic aspects will be discussed in another book [1].

---

[1]   J. D. Fast, *Interaction of Metals and Gases*, II *Kinetics and Mechanisms*, to be published.

CHAPTER 1

THERMODYNAMIC INTRODUCTION

This chapter gives a simple introduction to thermodynamics, which has not been carried further than is necessary for the purpose of this book. It will serve to refresh the memory of those readers who have already had some dealings with thermodynamics, but can also be understood by those to which this does not apply, provided that they have some elementary knowledge of physics, chemistry and higher mathematics.

In thermodynamics everything centres on the two laws and, consequently, as will be seen from the following, on the concepts of energy and entropy. Of the two, it is the latter which forms a stumbling block for many people, probably due to the abstract manner in which it is defined in classical thermodynamics. In the following we shall turn our attention to the statistical background of the entropy concept because only then does it take on a meaning as easily grasped as that of the energy concept [1]).

1.1. The two laws in classical thermodynamics

The nucleus of thermodynamics is formed by the two laws which can be expressed in many different ways. Superficially, these different formulations seem to have little in common, but in essence they are equivalent. Applied to an isolated system, i.e. a system which can not interact with the rest of the world, they can be expressed, for example, as follows. First law: The energy of an isolated system is constant. Second law: The entropy of an isolated system tends to a maximum.

The first law, also known as the law of conservation of energy, finds its origin in the experience that heat and work are mutually convertible forms of energy. If a system is not isolated (take, for example, a quantity of gas in a cylinder enclosed by a movable piston), then we can introduce a quantity

---

[1]) For a more thorough and less superficial discussion of the entropy concept the reader is referred to the book *Entropy* by the same author. In this book the applications of the entropy concept in many fields of science and technology are also discussed.

of heat dQ and perform work dW on it. According to the first law, both these quantities of energy will increase the internal energy U of the system:

$$dU = dQ + dW. \qquad (1.1.1)$$

Strictly speaking, until the subject of atoms has been broached, the concept of internal energy only acquires a clearly defined meaning from this formulation of the first law. In the atomic picture, of which classical thermodynamics is independent, the internal energy is the sum of the kinetic and potential energies of all the elementary particles of which the system is composed.

> In accordance with international agreements, we regard dQ as the heat introduced into a system and dW as the work performed on the system. On the basis of this convention, the heat absorbed by the system and the work performed on the system are always counted positive and thus the heat given up to the surroundings and the work done on the surroundings negative.

The internal energy depends only on the state of the system, i.e. on th-pressure, temperature, volume, chemical constitution, structure, etc. The previous history has no effect on its value. U is therefore called a thermoe dynamic function. W and Q are *not* thermodynamic functions, for, according to equation (1.1.1), the same change dU can be produced in the internal energy of a system either by the addition of heat alone or by performing work alone or by a combination of the two. It is therefore possible to speak of the internal energy of a system, but not of the quantity of heat or the quantity of work in that system. In other words: dW and dQ are only infinitesimal quantities of work and heat, they are not differentials of thermodynamic functions.

Although W is not a thermodynamic function and dW is not a differential, dW can in general be written as the product of an intensive thermodynamic function and the differential of an extensive thermodynamic function. Here the term "extensive quantity" indicates a quantity which is proportional to the extent of the system concerned, and "intensive quantity" a quantity which is independent of it. Thus for the volume-work already considered we can write:

$$dW = -pdV,$$

where p, the gas pressure, is the intensive thermodynamic function and V, the volume, is the extensive thermodynamic function. As long as p represents the internal pressure, this formula is only valid for a reversible change of volume, i.e. a change which takes place in such a way that the external pressure never differs by more than an extremely small amount from the internal pressure.

In an analogous manner dQ can be written

$$dQ = TdS,$$

where T, the temperature of the system, is the intensive and S, the entropy, is the extensive quantity. This formula, too, is only valid when heat is added reversibly, i.e. added from a reservoir of which the temperature is only extremely little above that of the gas. We thus have

$$dS = \frac{dQ_{rev}}{T}. \tag{1.1.2}$$

This formula not only gives the classical definition of the thermodynamic function S — and strictly speaking also of the absolute temperature T — but furthermore it gives the (still incomplete) mathematical formulation of the second law of thermodynamics.

For the reader who is unfamiliar with thermodynamics, the preceding sentences will not have clarified the entropy concept at all. He will also fail to see the connection between the given formulation and the above-mentioned "tendency of the entropy to a maximum". The simplest way to make this matter less obscure is to leave the purely thermodynamic path and make use of atomic considerations. Only when the atomic significance of the entropy concept has been explained, shall we return to the thermodynamic definition (1.1.2) of entropy and show how it must be modified in order to express the "tendency to a maximum".

1.2. The atomic significance of the entropy concept

IRREVERSIBLE PROCESSES

The second law originates in the experience that all processes occurring spontaneously have a definite direction, i.e. are irreversible. If the isolated system previously discussed consists, for example, of 0.5 gram-atom of neon and 0.5 gram-atom of helium enclosed in a container under such conditions that they may be regarded as perfect gases, then the *first* law permits all conceivable spatial arrangements of the gas molecules: the energy of the system is not influenced in any way if differences of concentration or possibly of pressure and temperature occur between different parts of the mixture. Experience shows, however, that whatever the initial state may be, the final state which the system assumes if left to itself (the "equilibrium state"), will always be that in which the gases are homogeneously mixed and the pressure and temperature are the same at all points. Having reached this final state, the system never returns spontaneously to one of the previous states.

What is the origin of this tendency to homogeneous mixing? In order to be able to answer this question, we shall first consider a simpler mixing experiment, carried out with a small number of white and red billiard balls (e.g. 50 + 50). For the initial condition we choose a certain regular arrangement of these balls in a container (Fig. 1) and we imitate the thermal agitation

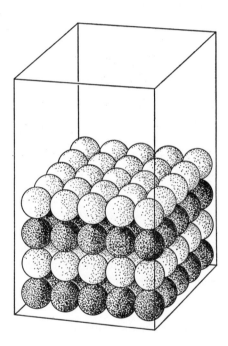

Fig. 1. A regular arrangement of fifty white and fifty red billiard balls.

of the atoms by thoroughly shaking the container for a certain length of time. Experience shows that after the container has been shaken, the regular initial arrangement is never encountered again but that the red and white balls always form a disordered distribution. Even so, we must assume that there is no "preference" for particular distributions, in other words that each separate disordered distribution is just as probable or improbable as the initial distribution. The point is, however — and here we touch the very core of the problem — that there are so many more irregular than regular configurations possible that virtually only irregular ones will occur in the shaking experiments. This will be illustrated with the help of a simple calculation.

A STATISTICAL CALCULATION

Suppose it is possible to distinguish between all of the 100 balls (e.g. by numbering them), then we can place them in $100 \times 99 \times 98 \times \ldots \times 2 \times 1 = 100!$ (factorial 100) different ways in the 100 available positions. Since, in reality, the 50 red balls can not be distinguished from one another (are not numbered), the reversal of any two red balls does not change the distribution. The number of arrangements m which can be distinguished by the eye is thus much smaller than 100! Nevertheless, this number is still extremely large. Since the white balls also are indistinguishable from one another, the number is given by

$$m = \frac{100!}{50!\,50!}.$$

Accurate values of the factorials of numbers from 1 to 100 are to be found in Barlow's Tables [1]. We find that $100! = 9.333 \times 10^{157}$ and $50! = 3.041 \times 10^{64}$.

From this follows the value for m

$$m = 1.01 \times 10^{29}.$$

A reasonably accurate approximation formula for calculating factorials is Stirling's formula:

$$N! = \frac{N^N}{e^N}\sqrt{2\pi N}, \tag{1.2.1}$$

where e is the base of natural logarithms. It can easily be verified that this equation gives the value of m quoted above. On the other hand, the number of balls chosen is too small to allow application of Stirling's formula in its roughest approximation

$$\ln N! \simeq N \ln N - N. \tag{1.2.2}$$

This gives for the value of m

$$m = 2^{100} = 1.26 \times 10^{30},$$

i.e. a value which is too large by a factor of more than 12. It is immediately obvious that the value 2^{100} not only comprises all arrangements of 50 white and 50 red balls, but also all arrangements of 100 balls without any restriction being made on the ratio of white to red, thus also all arrangements of 49 white and 51 red, of 48 white and 52 red, etc. When there is no restriction on the ratio white: red, there is a choice of two possibilities (white or red) to fill the first position. When two positions are to be filled, each of the two

---

[1] *Barlow's Tables*, E. & F. N. Spon Ltd., London, 1947.

colours can be combined with each of the two colours, so that there are 2^2 possibilities (w-w, w-r, r-w and r-r), etc.

As we have already said, we must assume that in our shaking experiments all $m = 10^{29}$ configurations of the balls (all "micro-states" or "possibilities of realization") have the same probability p:

$$p = \frac{1}{m}. \tag{1.2.3}$$

Thus it follows that one must shake an average of 10^{29} times in order to obtain a particular arrangement of the balls. If one only shakes for one minute each time, this means that an average of more than 10^{23} years shaking will be required before a particular distribution occurs. The above-mentioned experience is thus comprehensible: the chance that after shaking a completely regular arrangement will be found, e.g. one in which the white and red balls are present in separate layers, is virtually nil, since the number of regular arrangements is negligibly small compared with the enormous number of irregular arrangements.

In the mixture of helium and neon considered earlier, the continual regrouping is not brought about by "shaking" but by the thermal agitation of the particles. As we shall see, the chance of occurrence of a regular distribution of the many particles is incomprehensibly smaller than in the case of the hundred balls.

PROBABILITY OF GROUPS OF MICRO-STATES

The various micro-states can be grouped together according to their particular degree of disorder. An orderly distribution, as shown in Fig. 1, is only possible in one way. The same is true for the checkboard configuration in which each red ball has only white and each white one only red neighbours. In an orderly distribution an "imperfection" can be introduced by interchanging a white and a red ball. Since each of the 50 white balls can be interchanged with each of the 50 red ones, the group characterized by one imperfection contains 2500 micro-states. If one increases the number of imperfections to two or more, one obtains groups which contain considerably more micro-states. The probability p of each group is determined by the number of micro-states g which it comprises:

$$p = \frac{g}{m}. \tag{1.2.4}$$

The probability that a regular distribution with one imperfection will occur is therefore 2500 times as large as that of any arbitrary micro-state.

> Compare this with the well-known fact that the probability of throwing 7 points with two normal dice is six times as great as the probability of throwing 12 points. In the first case there are 6 configurations which produce the required number of points, viz. (1 & 6), (2 & 5), (3 & 4), (4 & 3), (5 & 2), (6 & 1), while in the latter case there is only one configuration, viz. (6 & 6).

The mixing experiment described above was carried out with the small number of 100 balls. On the other hand, the container mentioned at the beginning of the section contained $N_0 = 0.6 \times 10^{24}$ atoms, viz. 0.5 gram-atom helium and 0.5 gram-atom neon. By the same reasoning as above the number of possible arrangements of this number of "white" and "red" balls is given by

$$m = \frac{N_0!}{\{(\tfrac{1}{2}N_0)!\}^2} = 2^{N_0-40}. \tag{1.2.5}$$

The figure 40 in the exponent of 2 is quite negligible with respect to $N_0 = 0.6 \times 10^{24}$; Avogadro's number N_0 is known to so few decimals that it would actually be physically meaningless to try and make a distinction between N_0 and $N_0 - 40$. We may thus write

$$m \simeq 2^{N_0}$$

and this, from the above, is none other than the total number of arrangements of N_0 balls of two sorts. In other words: for large values of N the number of arrangements as a function of the "mixing ratio" shows such a sharp maximum at the ratio 1 : 1 that, in practice, it makes no difference whether one is dealing with the number of arrangements corresponding to this maximum or with the total number. It is very important that the approximation formula (1.2.2) also gives the result $m = 2^{N_0}$. For the numbers of atoms which are always dealt with in practice this formula is thus completely satisfactory.

The number 2^{N_0} is so astronomically large that it is quite beyond the grasp of the human imagination. As in the example with the small numbers, it must be assumed that all these configurations have the same probability. This, in turn, means that a given state is more probable when it can be realized by a greater number of configurations. This also means that the probability of an orderly distribution is incomprehensibly smaller than in the case of the hundred balls.

Due to the thermal agitation of the molecules, a mixture of gases passes spontaneously through one configuration after another. The fact that, even so, one always finds a state in which the gases (macroscopically seen) are homogeneously mixed, is not surprising in view of the above. In fact, this state comprises so very many more configurations than all other states together that after a short interval it is always observed to the exclusion of all others. It is called the thermodynamic equilibrium state because it always establishes itself spontaneously whatever the initial distribution may have been. What is known in classical thermodynamics as the tendency of an isolated system towards a maximum value of the entropy, corresponds in the atomic picture to the tendency towards a state with a maximum number of configurations (micro-states), i.e. to the tendency towards the most probable state.

THE STATISTICAL DEFINITION OF ENTROPY

If the entropy of a system increases, then, from the above, there will also be an increase in the number of atomic configurations in which the system can occur. The statistical definition of entropy which is based on this relationship takes the following form:

$$S = k \ln g, \tag{1.2.6}$$

where g is the number of configurations in equation (1.2.4), corresponding to the state with entropy S, and k is Boltzmann's constant, given by

$$k = \frac{R}{N_0}. \tag{1.2.7}$$

(R = gas constant, N_0 = Avogadro's number).

As we have seen, there is no objection to replacing g for the most probable state (g_{max}) by m, the total number of micro-states, and we may therefore write:

$$S = k \ln m. \tag{1.2.8}$$

The calculation of entropies by statistical means thus involves counting the numbers of configurations (possibilities of realization). To illustrate this we shall now calculate the increase in entropy which occurs during the isothermal expansion of a perfect gas.

We imagine that the volume containing the gas is divided into very many equal cells which are so small that by far the greater part are empty, while a minute fraction of the cells contain one gas molecule. As a result of the thermal agitation the occupied cells are continually changing. If there are z cells and N molecules, there will be N occupied and $(z - N)$ unoccupied cells. Since the interchange of either two empty or two occupied cells does not alter an arrangement, the total number of configurations is given by

$$m = \frac{z!}{N!(z - N)!} \qquad (1.2.9)$$

or, employing equation (1.2.2):

$$\ln m = z \ln z - N \ln N - (z - N) \ln (z - N).$$

Making use of the approximation $\ln (1 - N/z) = -N/z$ for $N/z \ll 1$ we get:

$$\ln m = N \ln \frac{z}{N} + N. \qquad (1.2.10)$$

If we increase the volume by a factor $r = V_2/V_1$, then z must be replaced in the equation by rz. The increase in entropy, according to (1.2.8) is thus given by

$$\Delta S = k \ln m_2 - k \ln m_1 = kN \ln \frac{rz}{N} - kN \ln \frac{z}{N} =$$

$$= kN \ln r = kN \ln \frac{V_2}{V_1}.$$

Thus for 1 gram-molecule (1 mole) gas, using (1.2.7):

$$\Delta S = R \ln \frac{V_2}{V_1}. \qquad (1.2.11)$$

We have already mentioned two possible causes of an increase in the number of configurations and hence in the entropy of a system: (1) mixing of particles of different kinds, (2) an increase in the volume in which the particles are free to move. A third cause for an increase in entropy is the absorption of thermal energy which can be distributed among the particles of a system in very many different ways. The relationship between the increase in entropy and the number of possible distributions is given also in this case by equation (1.2.6) or (1.2.8).

1.3. The two definitions of entropy and the justification of the formulae given

The formulae (1.1.2) and (1.2.8) have introduced us to two definitions of

entropy which, at first sight, have little or nothing in common and even seem
to lead to conflicting conclusions. If, for example, we double the volume of a
mole of a perfect gas by connecting the vessel containing it to an evacuated
container of the same size, the entropy according to (1.2.11) will increase,
but no heat will flow in or out of the system. At first sight one might suppose
that equation (1.1.2) can be applied and that the change in entropy is zero.

On further consideration, however, it is clear that equation (1.1.2) can not
be applied directly to this typically irreversible process. The formula is, in
fact, only valid for reversible processes and therefore, to calculate the change
in entropy, we must find a reversible path leading from the same initial state
to the same final state. One such path is the following. The gas is introduced
into a cylinder fitted with a frictionless movable (weightless) piston and the
expansion is carried out in surroundings at constant temperature in such a
way that at all times the pressure applied to the piston differs only infinite-
simally from the gas pressure. In this case the process is called reversible
because at every stage a very small change in the piston pressure is sufficient
to reverse the direction of the process.

During the reversible expansion the perfect gas performs work on the
surroundings, which is given by

$$\int dW = - \int p \, dV.$$

Since the internal energy of the gas is not changed by the isothermal
expansion, then, according to (1.1.1), a quantity of heat will be absorbed
from the surroundings, given by

$$\int dQ = + \int p \, dV.$$

From (1.1.2), the entropy change is given by

$$\Delta S = \int_{V_1}^{V_2} \frac{dQ}{T} = \int_{V_1}^{V_2} \frac{p \, dV}{T} = \int_{V_1}^{V_2} R \frac{dV}{V},$$

$$\Delta S = R \ln \frac{V_2}{V_1},$$

in complete agreement with (1.2.11).

The same entropy change must occur during *irreversible* expansion from
V_1 to V_2, because S is a thermodynamic function. During this irreversible
change of state, however, no heat is exchanged with the surroundings, so that

$$dS > \frac{dQ}{T},$$

and this is true, in general, for all irreversible processes. The second law can thus be written more generally than in (1.1.2), in the following form:

$$dS \geqslant \frac{dQ}{T}, \tag{1.3.1}$$

where the equality sign applies to a reversible change of state and the inequality sign to an irreversible change.

For an isolated system always $dQ = 0$ and thus from (1.3.1) $dS \geqslant 0$. Equation (1.3.1) therefore gives a mathematical formulation of the description given in Section 1.1 of the second law, according to which the entropy in an isolated system tends to a maximum. It was this very circumstance, viz. that an isolated system tends in the thermodynamic sense to a maximum value of its entropy and simultaneously in the atomic concept to a state with a maximum number of realization possibilities or configurations m, which suggested to Boltzmann in the last century the idea that there must be a functional connection between S and m. This connection was bound to be of a logarithmic nature, since entropy is an additive and the number of realization possibilities is a multiplicative quantity. That S is an additive quantity follows directly from the fact that for two systems $(A + B)$, separated but regarded as one system, the quantity of heat required to raise the temperature reversibly from T to $(T + dT)$ is the sum of the quantities of heat required to produce this temperature change in A and B separately (cf. equation (1.1.2)). Furthermore, if the numbers of realization possibilities for A and B are indicated by the symbols m_A and m_B, then this number will be given for $(A + B)$ by $m_A m_B$, since for *each* configuration of A, system B can occur in *each* of its m_B configurations. Once it was realized that there must be a logarithmic relationship between S and m, the value of the proportionality constant k in (1.2.8) was found by application to a perfect gas as described.

In the foregoing we have not gone into the historical developments which led to the mathematical formulations

$$dU = dQ + dW, \tag{1.1.1}$$

and

$$TdS \geqslant dQ \tag{1.3.1}$$

of the two laws of thermodynamics. At the present time, the most satisfactory course, logically, is to postulate the validity of these relationships and to see their justification in the fact that all conclusions deduced from them have been found in accordance with experiment. If one were ever able to contrive and carry out an experiment which contradicted either one of the two laws of thermodynamics, then the whole proud structure of thermodynamics would crumble and fall. Also the absolute temperature T introduced in (1.3.1) is justified in an analogous manner; it is found to be identical with the normal absolute temperature as measured by a gas thermometer.

The statistical definition of entropy, (1.2.6) or (1.2.8), is justified by the fact that in every case so far investigated it has led to the same results as the thermodynamic definition (1.1.2). We have only demonstrated this with one example, viz. the example of a perfect gas which expanded isothermally from V_1 to V_2.

1.4. Free energy and free enthalpy

In order to explain the significance of the entropy concept, we have restricted ourselves in the mixing experiments to a discussion of the perfect gaseous state, i.e. a state in which the molecules of the material exert no attractive forces on one another. In this state the equilibrium condition always corresponds to a disordered distribution of the various kinds of molecules. This is not necessarily the case if there are attractive forces between the molecules or if their distribution is influenced by external forces or fields. If, in the previously discussed case of the 100 billiard balls, the 50 white balls attracted each other strongly (e.g. by means of built-in magnets), then at the end of a shaking experiment one would generally find a distribution in which the system was separated into two "phases", one containing exclusively (or nearly so) white and the other exclusively (or nearly so) red balls. A disordered distribution of the balls now corresponds to a state of higher energy. This example shows that the striving towards a maximum value of the entropy may in some cases come into conflict with another tendency, namely that towards a minimum value of the energy. In order to find a quantitative relationship between the two, we combine equations (1.1.1) and (1.3.1) to give the equation

$$dU - TdS \leqslant dW, \tag{1.4.1}$$

or, for a constant value of T:

$$d(U - TS)_T \leqslant dW. \tag{1.4.2}$$

If, during the change of state, not only the temperature remains constant but also the volume (and, furthermore, all other parameters the variation of which can lead to the performance of external work), then $dW = 0$, so that for an *irreversible* process (1.4.2) becomes

$$d(U - TS)_{T,V} < 0. \tag{1.4.3}$$

Consequently, if we are not dealing with an isolated system, but with a system the temperature T and the volume V of which are kept constant, then the "tendency" of the entropy to a maximum is replaced by another tendency, namely that of the function $(U - TS)$ to a minimum. This thermodynamic function is called the Helmholtz free energy or free energy at constant volume and is indicated by the symbol F.

If, instead of the temperature and the volume, the temperature and the pressure are kept constant, then even for an irreversible process the system will still perform some work, viz. the volume-work $-dW = pdV$ (where p this time represents the external, not the internal, pressure), and it is immediately obvious from (1.4.2) that now another thermodynamic function, the function $U - TS + pV$, tends towards a minimum value. This function is known as the Gibbs free energy, but also the free energy at constant pressure or thermodynamic potential. It is generally indicated by the symbol G. Another name often used for this function is free enthalpy, which is derived from the name enthalpy for the thermodynamic function $H = U + pV$. If, in agreement with fairly general practice, the name Helmholtz free energy, $F = U - TS$, is abbreviated to free energy, then it is logical to call the function $G = H - TS$ the free enthalpy. For the sake of clarity, the three thermodynamic functions, which have been introduced by definition, are listed once more below [1]:

$$F = U - TS, \quad \text{(Helmholtz) free energy,}$$
$$H = U + pV, \quad \text{enthalpy,}$$
$$G = H - TS, \quad \text{free enthalpy or Gibbs free energy.}$$

The relationships found for irreversible processes

$$dF_{T,V} = d(U - TS)_{T,V} < 0 \tag{1.4.4}$$

and

$$dG_{T,p} = d(H - TS)_{T,p} < 0 \tag{1.4.5}$$

can be split, somewhat artificially, into two parts, $dU_{T,V} < 0$ or $dH_{T,p} < 0$;

---

[1] Warning: In the literature the symbol F is sometimes used instead of G for the Gibbs free energy. It is therefore always necessary to check the context carefully to be sure which of the two free energies is meant.

and $dS_{T,V} > 0$ or $dS_{T,p} > 0$. They can be regarded as the mathematical formulation of the two opposed tendencies which were indicated at the beginning of this section. If only the latter tendency ($dS > 0$) existed, one would expect only those processes and reactions to occur spontaneously for which the number of realization possibilities (the "disorder") increases. On the other hand, if only the first tendency were active ($dU < 0$ or $dH < 0$), one would only observe processes for which the opposite is the case, i.e. in which heat is liberated and in which, in general, the order increases. Thus, in order to determine the direction in which a process will take place we must consider the free energy, which involves both thermodynamic functions U and S, or H and S. If we do this we see from (1.4.4) and (1.4.5) that at low temperatures the tendency towards minimum energy (enthalpy) and the corresponding order predominates, while at high temperatures (violent shaking of the billiard balls) the tendency to maximum entropy and the accompanying disorder is predominant.

Accordingly, it is found that at low temperatures atoms and molecules, under the influence of their mutual attraction, form the orderly periodic structures we know as crystals. On the other hand, at high temperatures all matter is finally transformed into the chaotic gaseous state, often by way of the liquid state. The temperature range in which this entropy effect begins to predominate depends on the magnitude of the attractive forces. Even in gases a certain order is often present in the form of orderly groups of atoms (molecules). A further rise in the temperature, however, causes this order also to disappear and at the surface of our sun (temperature about 6000 °C) matter only exists as a gaseous mixture of the atoms of the various elements. But even this is not complete chaos and as the temperature rises higher and higher even the order in the electron shells is finally completely destroyed by thermal ionization. At a few million degrees, ionization is complete. This complete dissociation of atoms into nuclei and electrons prevails in the interior of our sun and of innumerable other stars. The only order remaining there is that of the nuclei. Complete dissociation of the nuclei into protons and neutrons would require higher temperatures than appear to occur in the hottest stars.

1.5. Zero point entropy and the third law

After this short digression into the region of very high temperatures we shall now consider that of extremely low temperatures.

According to Nernst's heat theorem, the entropy of all systems in stable

or metastable equilibrium tends to zero as the absolute zero of temperature
is approached. From the atomic point of view this is immediately understandable, since at absolute zero temperature all substances which are in
stable or metastable equilibrium are ideally crystallized and free of dissolved
components, while furthermore, they have not yet absorbed any energy
above the zero-point energy. Consequently, at that temperature they occur
in only one configuration: $m = 1$ and $S = k \ln 1 = 0$ [1]).

In many cases the validity of the theorem can be tested experimentally.
A classic example is the transition of white tin into grey tin. Grey tin is stable
below 13 °C (286 °K), white tin is stable above this temperature. Since white
tin can be supercooled to the lowest attainable temperatures, it has been
possible to measure the specific heat c_p of both modifications at low temperatures. If Nernst's theorem is correct, then from equation (1.1.2) the
entropy of white tin at 286 °K can be found in two ways, either directly
from c_p measurements on white tin or from those on grey tin and from the
entropy change which occurs with the transition from grey to white tin.
Both methods should lead to the same result, i.e. the following should be
valid:

$$\int_0^{286} \frac{c_p(w)\mathrm{d}T}{T} = \int_0^{286} \frac{c_p(g)\mathrm{d}T}{T} + \frac{Q}{286},$$

where $c_p(w)$ and $c_p(g)$ are the specific heats per gram-atom of white and grey
tin and Q the heat of transition, i.e. the quantity of heat absorbed during the
isothermal and reversible transition of one gram-atom of tin from the grey
to the white modification. It has, indeed, been found that the equation is
satisfied within the limits of experimental accuracy. Unfortunately, the heat
of transition Q is not known with such accuracy that great value can be
attached to this agreement. Moreover, even if complete agreement were
established this would only prove that the *difference* in entropy between the
two modifications at 0 °K is zero. The heat theorem is therefore often worded
in the following rather more cautious form: at the absolute zero-point all
entropy differences between the thermodynamical states of a system in
internal equilibrium vanish. This formulation has the same practical significance as that in which the separate entropies are said to approach zero, since
one can now usefully define the zero point of the entropy of all substances
in stable or metastable equilibrium as lying at absolute zero temperature.

---

[1]) For a more exact treatment of this question the reader is referred to an article by
H. B. G. *Casimir*, Z. Physik **171**, 246 (1963).

If for example, in every chemical reaction $A + B \rightleftharpoons AB$, the entropy change is zero at 0 °K, then it is logical to assign a zero-point entropy of zero to A and B as well as to AB. This is not possible for the energy, as extrapolations to absolute zero show that there is no question of the heats of reaction disappearing at the absolute zero-point.

Stronger evidence of the validity of Nernst's theorem is derived from measurements on gases. The statistical entropy formula (1.2.8), $S = k \ln m$, can be used to calculate the entropy of many gases, with the help of information derived from spectra on the rotational and vibrational states of their molecules. On the other hand, if Nernst's theorem is valid, the entropy of gases may be calculated with the help of the classical formula (1.1.2), $dS = dQ_{rev}/T$, making use of the known values of the specific heats c of these substances in the solid, liquid and gaseous states, and of the heats of transformation, fusion and evaporation. The entropy of a substance in the gaseous state at temperature T, assuming there are no transition points in the solid state, can be written directly as:

$$S = \int_0^{T_f} \frac{c_{sol}}{T}\, dT + \frac{Q_f}{T_f} + \int_{T_f}^{T_v} \frac{c_{liq}}{T}\, dT + \frac{Q_v}{T_v} + \int_{T_v}^{T} \frac{c_{gas}}{T}\, dT, \qquad (1.5.1)$$

where T_f and T_v are the melting and boiling points and Q_f and Q_v the heats of fusion and evaporation.

This "calorimetric entropy" is thus obtained entirely without reference to the existence of atoms; it is based solely on the results of calorimetric measurements. The "statistical entropy" on the other hand is found by a method which requires no knowledge of the existence of the liquid and solid states. The beauty of it is that both ways generally lead to the same result, while the exceptions which have been found can be satisfactorily explained.

Nernst's theorem cannot be derived from the first and second laws of thermodynamics and is therefore often called the "third law of thermodynamics".

CHAPTER 2

CHEMICAL EQUILIBRIUM

2.1. Chemical affinity

If a chemical reaction takes place in a system, volume-work is usually the only form of work which can be exchanged with the surroundings. Naturally, this volume-work will be zero if the volume of the system is kept constant. If, furthermore, the temperature is kept constant, then, according to the previous chapter (see equation (1.4.4)), the system will strive towards a minimum value of the free energy (or Helmholtz free energy)

$$F = U - TS.$$

On the other hand, if the temperature and *pressure* are kept constant, the free enthalpy (or Gibbs free energy)

$$G = U + pV - TS = H - TS$$

will tend to a minimum value, according to equation (1.4.5). In chemistry and metallurgy one is usually more interested, for technical reasons, in processes which take place at constant pressure than in those at constant volume. This implies a greater interest in the free enthalpy G than in the free energy F. The "competition" between U and S, the energy and the entropy, which occurs at constant volume, is replaced at constant pressure by that between H and S, the enthalpy and the entropy, viz. the tendency of H to a minimum and of S to a maximum value (see Section 1.4).

At constant temperature and pressure, according to equation (1.4.5), the only chemical reactions which will take place spontaneously will be those in which the change in the free enthalpy

$$\varDelta G = \varDelta H - T\varDelta S \qquad (2.1.1)$$

is negative. In this equation $\varDelta S$ is the increase in entropy caused by the reaction, while $\varDelta H$ is the heat of reaction at constant pressure, i.e. the quantity of heat absorbed by the system when the reaction takes place irreversibly at constant values of p and T. This follows directly from the first law (equation (1.1.1)) which here takes the form

$$Q_{irr} = \Delta U + p\Delta V = \Delta H. \qquad (2.1.2)$$

It may be useful to recall that the heat introduced into a system is regarded as positive (see Section 1.1), so that an exothermic reaction has a negative heat of reaction.

ΔU and ΔH, the heats of reaction at constant volume and constant pressure, are virtually equal when no gaseous substances appear in the formula for the reaction (e.g. $PbS + Fe \rightarrow Pb + FeS$) or when the number of gas molecules remains unchanged by the reaction (e.g. $H_2 + Cl_2 \rightarrow 2\,HCl$). If, however, the number of gas molecules formed during the reaction is not equal to the number which disappear (as in the reaction $2\,NH_3 \rightarrow N_2 + 3\,H_2$), then ΔU and ΔH will differ, according to (2.1.2), by an amount

$$p\Delta V = \Delta(pV) \cong nRT,$$

where n is the number of moles of gas formed less the number which disappears.

From the thermodynamic viewpoint, a chemical reaction will have a greater tendency to occur as ΔG for this reaction becomes more negative. Therefore, in thermodynamics $-\Delta G = -\Delta H + T\Delta S$ is often called the *affinity* of the reaction. It increases as ΔH becomes more strongly negative, i.e. the reaction more exothermic, and as ΔS becomes more strongly positive, i.e. the greater the increase in entropy during the reaction.

Chemical reactions in which only solids are involved are usually accompanied by a comparatively small change in entropy and are thus generally exothermic when they occur spontaneously. When gases appear in the reaction formula, ΔS will generally be small if the number of gas molecules remains unchanged during the reaction. If, on the other hand, the number of gas molecules increases in the course of the reaction, the entropy will generally increase considerably, since the number of available micro-states m is much greater in the gaseous state than in the condensed state (see Chapter 1, where we saw that $S = k \ln m$).

As an example, let us consider the oxidation of carbon according to the two equations

$$C + O_2 \rightarrow CO_2, \qquad (2.1.3)$$

$$2\,C + O_2 \rightarrow 2\,CO. \qquad (2.1.4)$$

During the course of (2.1.3) much more heat is evolved (the enthalpy decreases much more) than during the course of (2.1.4). On the other hand the entropy hardly changes during (2.1.3), while in (2.1.4), due to the doubling of the number of gas molecules, it increases considerably. If only a tendency towards minimum enthalpy existed, all carbon would oxidize to CO_2. If there were only the tendency towards maximum entropy, combustion would only produce CO. In reality a compromise is reached by the formation of a mixture of the two gases. At relatively low temperatures the energy (or enthalpy) effect predominates and carbon is chiefly converted into CO_2. At high temperatures, multiplication by T (see equation (2.1.1)) makes the entropy term more important and CO is principally produced.

In the latter case (ΔS positive), according to (2.1.1), ΔG will become more negative as the temperature rises. In other words: the affinity of carbon for oxygen, as regards the formation of CO, increases continuously with a rise of temperature. Conversely, in the case of oxidation of solid or liquid metals into solid or liquid oxides, e.g.

$$2\,Fe + O_2 \rightarrow 2\,FeO, \qquad (2.1.5)$$

$$\tfrac{4}{3}\,Al + O_2 \rightarrow \tfrac{2}{3}\,Al_2O_3, \qquad (2.1.6)$$

the affinity decreases with rising temperature, since ΔS is negative due to the decrease in the number of gas molecules. The fact that rising temperature causes a decline in the affinity of oxidation reactions of the type (2.1.5) or (2.1.6) and a simultaneous increase in the affinity of reaction (2.1.4) means that for every metal there is some temperature above which oxygen at a certain pressure has a smaller tendency to combine with it than with carbon. We have thus here the entropy effect which is so important in extractive metallurgy: at a sufficiently high temperature all liquid or solid metal oxides can be reduced by carbon.

For a more quantitative treatment of the above, we must know the dependence of ΔG on the gas pressure. This dependence can most easily be found by introducing *molar quantities*, each of which relates to 1 gram-molecule or gram-atom (1 mole) of a particular substance.

2.2. Molar and partial molar quantities

Let us consider a chemical reaction taking place at constant values of p and T, e.g. the reaction

$$PbO_2 + 2\,Fe \rightarrow Pb + 2\,FeO.$$

The system embracing the substances participating in the reaction suffers a change of volume ΔV, a change of entropy ΔS, a change of enthalpy ΔH and so on. These changes are given by the molar quantities. This is most easily demonstrated by first examining the volume; it is clear that the initial volume of the system is that of 1 mole of PbO_2 and 2 moles of Fe, the final volume is that of 1 mole Pb and 2 moles FeO. The total change of volume is thus given by

$$\Delta V = v_{Pb} + 2\, v_{FeO} - v_{PbO_2} - 2\, v_{Fe}.$$

In general:

$$\Delta V = \Sigma v_i v_i, \tag{2.2.1}$$

where the symbol v_i refers to the molar quantities under consideration, i.e. the molar volumes of the various substances, and v_i to the coefficients in the equation of the reaction. The latter are always regarded as positive for the substances produced during the reaction (the products) and negative for those which vanish (the reactants).

If the substances taking part in a reaction are in solution, we must make use of *partial molar quantities*, thus in the case under consideration of partial molar volumes. The partial molar volume of a component i in a solution is the change in volume which occurs when 1 mole of this component is added to a quantity of solution so large that its composition does not change noticeably. If the partial molar volume of i is indicated by the symbol $\bar{v}_i$, then by definition the following is valid for constant values of p and T:

$$\bar{v}_i = \left(\frac{\partial V}{\partial n_i}\right)_{p,T,n_j}. \tag{2.2.2}$$

Here V is the volume of the solution and n_i the number of moles of i; n_j refers to the numbers of moles of the other components in the solution. The partial molar volume, like the other partial molar quantities still to be discussed, is in general dependent on the concentration of the solution. When certain salts are dissolved in water a decrease in the volume occurs, so that the partial molar volume can also be negative.

With the foregoing considerations in mind it is evident that equation (2.2.1) may also be used when substances in solution take part in a reaction, provided that v_i is replaced by the partial molar volume $\bar{v}_i$.

What has been said here about volume changes which occur in the course of a reaction can be reiterated in almost exactly the same words for changes in H, S, G, etc. One obtains relationships completely analogous to (2.2.1), e.g.:

$$\Delta H = \Sigma v_i h_i \quad \text{or} \quad \Delta H = \Sigma v_i \bar{h}_i, \tag{2.2.3}$$

$$\Delta S = \Sigma v_i s_i \quad \text{or} \quad \Delta S = \Sigma v_i \bar{s}_i, \tag{2.2.4}$$

where h or $\bar{h}$ indicates the molar or partial molar enthalpy and s or $\bar{s}$ the molar or partial molar entropy. For historical reasons, the molar or partial molar free enthalpy is not indicated by the symbol g or $\bar{g}$, but by the symbol μ

$$\Delta G = \Sigma \nu_i \mu_i. \tag{2.2.5}$$

Gibbs introduced the partial molar free enthalpy by the name *chemical potential*. In analogy to (2.2.2) the following is valid by definition:

$$\mu_i = \left(\frac{\partial G}{\partial n_i}\right)_{p,T,n_j}. \tag{2.2.6}$$

Naturally, corresponding formulae are valid for $\bar{h}_i$ and $\bar{s}_i$.

Since only those reactions for which ΔG is negative can occur spontaneously, reactions for which ΔG is positive have a tendency to proceed in the reverse direction. When $\Delta G = 0$ holds for a chemical reaction, it cannot proceed in either direction. In such a case, the chemical system in question is in equilibrium. Proceeding from (2.2.5), we can thus write as a condition of chemical equilibrium

$$\Delta G = \Sigma \nu_i \mu_i = 0. \tag{2.2.7}$$

As an example, we shall apply this formula to the case of steel, which is being melted under slag containing FeO. For the transition reaction

FeO (in the slag) $\leftrightharpoons$ FeO (in the steel)

the equilibrium condition is given directly by (2.2.7):

$$\mu_{\text{FeO}}^{\text{slag}} = \mu_{\text{FeO}}^{\text{metal}}. \tag{2.2.8}$$

The reaction thus comes to a halt as soon as the chemical potential of FeO has the same value in the two phases. Similar relationships are valid for other reactions.

The foregoing makes clear the significance of the name "chemical potential" for the partial molar free enthalpy. If a phase, containing a substance in solution, is brought into contact with another phase in which the substance has a lower potential, it will pass from the first to the latter phase until the potentials are equalized. Chemical potential thus plays a part in the distribution equilibrium which is analogous to that played by electrical potential in electrical equilibrium, by pressure in elastic equilibrium and by temperature in thermal equilibrium.

2.3. Dependence of chemical potential on pressure

According to the above, the chemical potential of a pure substance is nothing other than the molar free enthalpy of that substance, In that case, we have

$$\mu = u - Ts + pv, \tag{2.3.1}$$

where u is the molar energy, s the molar entropy and v the molar volume of the pure substance. From (2.3.1)

$$d\mu = du - Tds - sdT + pdv + vdp. \tag{2.3.2}$$

From the first and second laws the following is valid for our pure substance under equilibrium conditions:

$$du = Tds - pdv.$$

Substitution in (2.3.2) gives:

$$d\mu = -sdT + vdp. \tag{2.3.3}$$

At constant temperature, therefore,

$$d\mu = vdp$$

or, integrating:

$$\mu = \mu^0 + \int_{p=1}^{p} vdp, \tag{2.3.4}$$

where μ^0 is the molar free enthalpy (the chemical potential) of the pure substance under discussion at the temperature chosen and at unit pressure. As unit of pressure we choose 1 atmosphere; this choice will be adhered to in all further thermodynamic discussions. If we apply equation (2.3.4) to a perfect gas, we must substitute the value RT/p for v. We then obtain

$$\mu = \mu^0 + RT \ln p. \tag{2.3.5}$$

This relationship is also valid for the *partial* molar free enthalpy (the chemical potential) of a gas in a mixture of perfect gases, provided that its partial pressure is inserted for p. The reason for this is that no forces are active between the molecules in a mixture of perfect gases, so that the properties of each component in the mixture are the same as if it alone were present in the available space. Gases at a pressure of the order of 1 atmosphere, at temperatures well above their boiling point can be regarded, by approximation, as being perfect.

For a condensed phase (liquid or solid), the value of the integral in (2.3.4) will be negligibly small provided that the pressure changes are limited between zero and a few atmospheres. The molar volume of a solid or liquid phase in that region is virtually independent of the pressure. For example, if it is 20 cm³ then for $p = 0.01$ atm the absolute value of the integral will be 0.99×20 cm³.atm, which is equivalent to only about 0.5 cal. On the other hand, for a perfect gas the absolute value of the integral in this case is $RT \ln 100$, which corresponds to about 2700 cal at 300 °K. If the pressure changes do not exceed a few atmospheres, the chemical potential of a solid or liquid substance can thus be equated, to a good approximation, to the chemical potential at a pressure of 1 atmosphere:

$$\mu = \mu^0. \tag{2.3.6}$$

The principal reason for this is the small molar volume of a condensed phase.

2.4. Standard affinity and reaction constant

For chemical equilibrium, equation (2.2.7) is valid. If only gases and pure condensed substances take part in a reaction, the values of μ are given for the gases (at not too high pressures) approximately by (2.3.5) and for the condensed substances by (2.3.6). We thus obtain for ΔG:

$$\Delta G = \Sigma \nu_i \mu_i^0 + \Sigma \nu_j RT \ln p_j, \tag{2.4.1}$$

in which the terms $\nu_j RT \ln p_j$, only apply to the gases participating in the reaction, while the terms $\nu_i \mu_i^0$ apply to *all* the substances taking part. It is therefore convenient to write

$$\Sigma \nu_i \mu_i^0 = \Delta G^0, \tag{2.4.2}$$

in which ΔG^0 is the change in the free enthalpy which is caused by the reaction in question in the special case when all the substances involved are in their "standard state". In the cases which interest us here, the standard state is the pure state for liquids and solids and the perfect state at a pressure of 1 atm for gases. The quantity $-\Delta G^0$ is called the "standard affinity" of the reaction.

Combination of (2.4.1) and (2.4.2) gives:

$$\Delta G = \Delta G^0 + \Sigma RT \ln p_{j'}^{\nu_i}, \tag{2.4.3}$$

where the ν's are once again positive for the gaseous products and negative for the gaseous reactants. In the equilibrium state $\Delta G = 0$ and thus, for this state, we have:

$$\Delta G^0 = -RT \ln \Pi \, p_{j'}^{\nu_i}. \tag{2.4.4}$$

Here the sum of some logarithms has been replaced by the logarithm of a product (Π is the mathematical symbol for a continued product).

Since the quantities μ^0 are pure functions of the temperature, the right-hand side of (2.4.4) will vary only with the temperature. The product occurring there must therefore have a constant value at a given temperature. It is called the equilibrium constant or reaction constant K_p:

$$K_p = \Pi p_j^{\nu_j}. \qquad (2.4.5)$$

As seen in (2.4.4), the reaction constant is given by the standard affinity:

$$\Delta G^0 = -RT \ln K_p. \qquad (2.4.6)$$

Equation (2.4.6) is one of the most frequently used equations in "chemical thermodynamics".

For a simple oxidation reaction, e.g. (2.1.5) or (2.1.6), the value of K_p is found from (2.4.5) to be

$$K_p = p_{O_2}^{-1}.$$

In this case, from (2.4.6) we have

$$\Delta G^0 = RT \ln p_{O_2}, \qquad (2.4.7)$$

where p_{O_2} is the dissociation pressure of FeO or Al_2O_3. For reaction (2.1.4) it follows from (2.4.5) that

$$K_p = p_{CO}^2 \cdot p_{O_2}^{-1},$$

so that in this case ΔG^0 is given by

$$\Delta G^0 = -RT \ln \frac{p_{CO}^2}{p_{O_2}}. \qquad (2.4.8)$$

The equilibrium

$$2\,C + O_2 \rightleftharpoons 2\,CO$$

at any temperature can thus be calculated directly by means of (2.4.8) if the value of ΔG^0 for the reaction is known as a function of temperature.

2.5. Determination of ΔG^0 values

There are three methods by which the standard affinities of chemical reactions can be determined quantitatively. Sometimes they are determined directly from equilibrium measurements, i.e. from measurements of gas pressures in the case of reactions between gases and pure condensed substances (see above). Use is then made of the relationship, given above,

$$\Delta G_T^0 = -RT \ln K_p. \qquad (2.5.1)$$

In many cases ΔG^0 values can be obtained from calorimetric data by means of the relationship (2.1.1), applied to a reaction occurring under standard conditions:

$$\Delta G_T^0 = \Delta H_T^0 - T\Delta S_T^0. \qquad (2.5.2)$$

ΔS_T^0, the entropy change which occurs when the reaction takes place under standard conditions at a temperature T, can be calculated with the help of (2.2.4) if the molar entropies of the reacting substances are known. These can de calculated according to the "third law" if the molar specific heats of the reacting substances are known in the range between $0°K$ and $T °K$ and, furthermore, if one has knowledge of the latent heats of all physical changes of state undergone in this range by the substances participating in the reaction. For details of the necessary calculations, the reader is referred to Section 1.5 and, in particular, to equation (1.5.1), which may need to be extended to account for any crystallographic transformations which may occur. The molar entropies of many gases can also be derived from spectroscopic data. ΔH^0, the heat of reaction at constant pressure under standard conditions, is known, in many cases, at one temperature from calorimetric measurements. With the help of molar specific heats and latent heats it can then be calculated for the required temperature T (see Section 3.2).

There are some cases in which ΔG^0 can be measured directly as an electromotive force. To illustrate this, we shall consider a chemical reaction which can proceed in a galvanic cell, supplying electrical energy, e.g. the reaction

$$CuSO_4aq + Zn \rightarrow ZnSO_4aq + Cu \qquad (2.5.3)$$

or, abbreviated:

$$Cu^{++} + Zn \rightarrow Zn^{++} + Cu. \qquad (2.5.4)$$

In principle, a reaction of this sort can be made to proceed reversibly by applying an opposing voltage to the cell, which is only very slightly smaller than the electromotive force E of the cell. If the reaction proceeds until dn mole $CuSO_4$ has been consumed, a charge $zFdn$ will have flowed through the circuit, where F is the charge on one gramion of monovalent positive ions ($F = 1$ faraday $= 96500$ coulombs) and z is the number of faradays transported during the conversion of 1 mole $CuSO_4$ ($z = 2$ in this particular case). The electrical work performed is then

$$dW = -zFEdn. \qquad (2.5.5)$$

The negative sign is in accordance with the agreement made in Section 1.1, that dW shall represent the work *performed on the system*. The total amount of work performed on the system is

$$dW_{rev} = -pdV - zFEdn, \qquad (2.5.6)$$

where dV represents the increase in volume which accompanies the conversion of dn moles at the pressure p. Application of the first and second laws gives

$$dU = dQ_{rev} + dW_{rev},$$

$$dU = TdS - pdV - zFEdn. \qquad (2.5.7)$$

Since $G = U - TS + pV$, the following is valid at constant values of p and T:

$$dG = -zFEdn \qquad (2.5.8)$$

or, for the conversion of a quantity of matter as indicated by the reaction equation, e.g. (2.5.3) or (2.5.4):

$$\Delta G = -zFE.$$

At a particular temperature and under specially chosen standard conditions, this equation takes the form

$$\Delta G_T^0 = -zFE_T^0. \qquad (2.5.9)$$

Values of ΔG are usually given in kcal. In order to express the value of the right-hand side of the equation in kcal, use must be made of the relationship

$$1 \text{ faraday volt} = 23.06 \text{ kcal},$$

i.e. the value

$$F = 23.06 \text{ kcal/volt}$$

must be substituted for F in equation (2.5.9):

$$\Delta G_T^0 = -23.06 \, zE_T^0 \text{ kcal}. \qquad (2.5.10)$$

Summarizing, we note that ΔG^0 values can be determined with the help of the relationships (2.5.1), (2.5.2) and (2.5.9).

2.6. Rate and maximum yield of chemical reactions

The affinity values do not give us any information about the *rate* of a chemical reaction. Cases are known in which ΔG is strongly negative and yet the rate of reaction is negligibly small. For example, the standard affinity of reaction (2.1.6) at room temperature is very large, but from experience we know that aluminium objects do not react appreciably with oxygen at this temperature. Examination reveals that a very thin film of oxide is formed on the surface of the metal, protecting it from further corrosion, since neither aluminium nor oxygen atoms (or ions) can permeate through this skin.

Investigation of the rate of chemical reactions falls outside the scope of thermodynamics, which is only concerned with equilibrium states. The great value of thermodynamics in chemistry becomes clear, however, when one considers that it must first be known whether ΔG is negative for a desired reaction, before any useful purpose can be served by looking for suitable catalysts to accelerate the reaction. It is of even greater importance that from ΔG^0, as shown in preceding sections, one can calculate the chemical equilibrium, i.e. the *maximum yield* of the reaction products. Just how useful this can be to the metallurgist appears, for instance, from the striking example noted in the last century by the great French chemist Le Chatelier. He described it approximately as follows.

It is well-known that the reduction of iron oxides in blast-furnaces is mainly effected by their reaction with CO, forming CO_2. The gas leaving the blast-furnace at the top, however, contains a considerable percentage of CO, which was naturally considered wasteful and undesirable. Since the incompleteness of the reaction was ascribed to an insufficiently prolonged contact between CO and the iron ore, blast-furnaces of abnormal height were built. The percentage of CO in the escaping gases did not, however, diminish. These costly experiments showed that there is an upper limit to the power of CO to reduce iron ore, which cannot be exceeded. Familiarity with the thermodynamic laws of chemical equilibrium would have enabled the same conclusion to be reached much more quickly and at far less cost.

We might add to this that it is only necessary to know the reaction enthalpy ΔH^0 and the reaction entropy ΔS^0 of the reaction

$$FeO + CO \rightarrow Fe + CO_2 \qquad (2.6.1)$$

for the standard conditions and technically important temperatures in order to be able to calculate the maximum fraction of the CO which can be converted into CO_2. This calculation can be carried out directly with the help of the equations

$$\Delta G^0 = \Delta H^0 - T\Delta S^0$$

and

$$\ln \frac{p_{CO_2}}{p_{CO}} = -\frac{\Delta G^0}{RT}. \qquad (2.6.2)$$

The actual calculation will not be carried out until Chapter 4, after the introduction in the next chapter of a table containing the required thermo-dynamic data.

CALCULATION OF EQUILIBRIUM

3.1. Tabulated values of enthalpies of formation, free enthalpies of formation and molar entropies

When making thermodynamic calculations in chemistry and metallurgy, advantageous use can be made of tables like Table 1. The data appearing in this table refer to formation reactions in which a compound is formed from the elements. For these reactions the table gives the enthalpy of reaction ΔH^0 and the free enthalpy of reaction ΔG^0 at a temperature of 25 °C and the standard conditions discussed in Chapter 2. The table also gives the molar entropies of the tabulated substances, also at 25 °C and under standard conditions. These conditions specify that the pressure shall be 1 atm, that the condensed substances shall be pure and that the gases shall be in the perfect state.

It is assumed, in compiling the table, that the elements from which the various substances are formed (with the exception of phosphorus) are in the state which is most stable at 25 °C and 1 atm. Thus, for example, AgBr is assumed to be formed from solid silver and liquid bromine, Ag_2S from solid silver and rhombic sulphur. The state of any substance is indicated by the symbol s for the solid, l for the liquid and g for the gaseous state.

Table 1

Standard values at 25 °C of the enthalpy of formation ΔH^0 in kcal/mole, the free enthalpy of formation ΔG^0 in kcal/mole and the entropy s^0 in cal/deg.mole. The letters s, l and g indicate the solid, liquid and gaseous states. The letter s is replaced by a number or a Greek letter where it is necessary to specify more precisely which solid state is meant: 1) disthene, 2) sillimanite, 3) graphite, 4) diamond, 5) calcite, 6) aragonite, 7) wollastonite, 8) pseudowollastonite, 9) pyrite, 10) red, 11) black, 12) yellow, 13) white, 14) red, 15) red, 16) rhombic, 17) monoclinic, 18) quartz, 19) glass, 20) white, 21) grey, 22) rutile, 23) sphalerite, 24) wurtzite.

		ΔH^0	ΔG^0	s^0			ΔH^0	ΔG^0	s^0
Ag	s	0	0	10.20	$AgNO_3$	s	$-$ 29.4	$-$ 7.7	33.7
AgBr	s	$-$ 23.8	$-$ 22.4	25.6	Ag_2O	s	$-$ 7.3	$-$ 2.6	29.1
AgCN	s	34	39	20.0	Ag_2S	α	$-$ 6.6	$-$ 8.6	34.8
Ag_2CO_3	s	$-$121.0	$-$104.5	40.0	Ag_2S	β	$-$ 6.0	$-$ 8.4	35.9
AgCl	s	$-$ 30.36	$-$ 26.22	23.0	Ag_2SO_4	s	$-$170	$-$147	47.8
Ag_2CrO_4	s	$-$170	$-$148.5	51.8	Al	s	0	0	6.75
AgI	s	$-$ 14.9	$-$ 15.8	27.3	$AlBr_3$	s	$-$126	$-$121	

		ΔH^0	ΔG^0	s^0			ΔH^0	ΔG^0	s^0
Al_4C_3	s	− 47		31.3	$CaSiO_3$	8)	−377.4	−357.4	20.9
$AlCl_3$	s	−166	−152	26.3	Cd	s	0	0	12.3
AlN	s	− 76.5			Cd	g	27.0	18.7	40.1
Al_2O_3	α	−399	−377	12.2	$CdCO_3$	s	−180	−162	
$Al_2(SO_4)_3$	s	−821	−739	57.2	CdO	s	− 61	− 54	13.1
Al_2SiO_5	1)	−642.7	−607.0	20.7	CdS	s	− 34.5	− 33.6	17
Al_2SiO_5	2)	−648.9	−615.0	26	$CdSO_4$	s	−221.4	−196.0	29.4
As	s	0	0	8.4	Cl_2	g	0	0	53.28
Au	s	0	0	11.3	Co	s	0	0	7.0
Au_2O_3	s	19	39	30	CoAl	s	− 26		
B	s	0	0	1.4	Co_3C	s	9.5	7.1	29.8
BN	s	− 60.7		3.67	$CoCl_2$	s	− 77	− 67	25.4
B_4C	s	− 14		6.5	CoO	s	− 57.5	− 51	12.65
B_2O_3	s	−300		13	Cr	s	0	0	5.68
Ba	s	0	0	16	Cr_4C	s	− 16	− 17	25.3
$BaCO_3$	s	−291	−272	26.8	Cr_7C_3	s	− 43	− 44	48.0
BaH_2	s	− 41			CrN	s	− 29		
Ba_3N_2	s	− 90	− 73		Cr_2O_3	s	−269.7	−250.2	19.4
BaO	s	−133	−126	16.8	CrO_3	s	−138		
$BaSO_4$	s	−350	−323	32	Cs	s	0	0	21.2
Be	s	0	0	2.28	Cu	s	0	0	7.97
Be_3N_2	s	−135			$CuCO_3$	s	−142	−124	
BeO	s	−146	−139	3.4	CuO	s	− 37	− 30.5	10.4
$BeSO_4$	s	−286			Cu_2O	s	− 40	− 35.0	24
Bi	s	0	0	13.6	CuS	s	− 11.6	− 11.7	15.9
Bi_2O_3	s	−138	−119	36	Cu_2S	β	− 19.0	− 20.6	28.9
Br_2	l	0	0	36	$CuSO_4$	s	−184.0	−158.2	27
Br_2	g	7.4	0.75	58.64	F_2	g	0	0	48.6
C	3)	0	0	1.36	Fe	s	0	0	6.5
C	4)	0.45	0.68	0.58	$FeAl_3$	s	− 27		
CCl_4	l	− 33.3	− 16.4	51.3	Fe_3C	s	5	4	25.7
CCl_4	g	− 25.5	− 15.4	74.0	$FeCO_3$	s	−179	−161	22.2
CH_4	g	− 17.9	− 12.1	44.5	$FeCl_2$	s	− 81.5	− 72.2	28.6
C_2H_6	g	− 20.2	− 7.9	54.8	Fe_4N	s	− 1.1	2.4	37.3
C_2H_4	g	12.5	16.3	52.5	$Fe_{0,95}O$	s	− 64	− 58	13.4
C_2H_2	g	54.2	50.0	48.0	Fe_2O_3	s	−197	−177	21.5
C_6H_6	l	11.5	29.7	41.0	Fe_3O_4	s	−267	−242	35.0
C_6H_6	g	19.8	31.0	64.3	FeS	α	− 22.7	− 23.3	16.1
$(CN)_2$	g	74	71	57.9	FeS_2	9)	− 42.5	− 39.8	12.7
CO	g	− 26.4	− 32.8	47.30	$FeSO_4$	s	−220		25.7
CO_2	g	− 94.05	− 94.26	51.06	FeSi	s	− 19.2	− 19.5	12
CS_2	l	21.0	15.2	36.1	Fe_2SiO_4	s	−344	−320	35.4
CS_2	g	27.5	15.5	56.8	Ga	s	0	0	10.2
Ca	s	0	0	9.95	Ge	s	0	0	10.1
CaC_2	s	− 15	− 16	16.8	H_2	g	0	0	31.21
$CaCO_3$	5)	−288.4	−269.8	22.2	H	g	52.0	48.6	27.4
$CaCO_3$	6)	−288.5	−269.5	21.2	HBr	g	− 8.5	− 12.7	47.46
$CaCl_2$	s	−190.0	−179.3	27.2	HCN	g	31.2	28.7	48.2
CaF_2	s	−290.3	−277.7	16.5	HCl	g	− 21.9	− 22.7	44.64
CaH_2	s	− 45	− 36	10	HF	g	− 64.2	− 64.7	41.5
Ca_3N_2	s	−105	− 90		HI	g	6.20	0.31	49.34
CaO	s	−151.9	−144.4	9.5	HNO_3	l	− 41.4	− 19.1	37.2
$Ca_3(PO_4)_2$	s	−989	−932	56.4	H_2O	l	− 68.32	− 56.69	16.72
CaS	s	−115	−114	13.5	H_2O	g	− 57.80	− 54.63	45.11
$CaSO_4$	s	−342.4	−315.6	25	H_2O_2	l	− 44.8	− 27.7	
$CaSiO_3$	7)	−378.6	−358.2	19.6	H_2S	g	− 4.81	− 7.89	49.16

		ΔH^0	ΔG^0	s^0			ΔH^0	ΔG^0	s^0
H_2SO_4	l	−194			MnO_2	s	−125	−111	12.7
H_2Se	g	20.5	17.0	52.9	Mn_2O_3	s	−232		26.4
Hf	s	0	0	13.1	Mn_3O_4	s	−331	−306	35.5
HfO_2	s	−271			MnS	s	− 47		18.7
Hg	l	0	0	18.5	$MnSO_4$	s	−254	−228	
Hg	g	14.54	7.59	41.8	$MnSiO_3$	s	−302	−283	21
$HgCl_2$	s	− 53		35	Mo	s	0	0	6.8
$HgCl_2$	g	− 34	− 33.5	70	Mo_2C	s	4	3	19.7
Hg_2Cl_2	s	− 63.3	− 50.3	46.8	$MoCl_5$	s	− 91		
HgO	10)	− 21.7	− 14.0	17.2	Mo_2N	s	− 17	− 10	
HgS	10)	− 13.9	− 11.7	18.6	MoO_3	s	−180.3	−162	18.7
HgS	11)	− 12.9	− 11.05	19.9	MoS_2	s	− 55.5	− 54	15.1
Hg_2SO_4	s	−177.3	−149.1	48.0	MoS_3	s	− 61	− 57	16
In	s	0	0	13	N_2	g	0	0	45.77
Ir	s	0	0	8.7	N	g	85.6	81.5	36.61
I_2	s	0	0	27.9	NH_3	g	− 11.05	− 3.95	46.03
I_2	g	14.88	4.63	62.26	NH_4Cl	s	− 75.4	− 48.7	22.6
I	g	25.5	16.7	43.2	$(NH_4)_2SO_4$	s	−282	−215	53
K	s	0	0	15.2	NO	g	21.6	20.72	50.33
K	g	21.5	14.6	38.3	N_2O	g	19.5	24.8	52.6
K_2	g	30.8	22.1	59.7	NO_2	g	8.0	12.3	57.5
KBr	s	− 93.7	− 90.5	23	N_2O_4	g	2.3	23.5	72.7
K_2CO_3	s	−274			Na	s	0	0	12.2
KCl	s	−104.2	− 97.6	19.8	Na	g	26.0	18.7	36.7
$KClO_3$	s	− 94	− 70	34.2	Na_2	g	34.0	24.9	55.0
$KClO_4$	s	−104	− 73	36.1	NaBr	s	− 86.0		20
KF	s	−134.5	−127.4	15.9	Na_2CO_3	s	−270.3	−250.4	32.5
KH	s	− 15.0			NaCl	s	− 98.3	− 91.8	17.3
KI	s	− 78.3	− 77.0	24.9	NaF	s	−136	−129	12.3
KIO_3	s	−121.5	−101.7	36.2	NaH	s	− 13.6		
$KMnO_4$	s	−194	−171	41.0	NaI	s	− 68.9		22.5
KNO_3	s	−117.8	− 94.0	31.8	$NaNO_3$	s	−111.5	− 87.5	27.8
K_2SO_4	s	−342.6	−314.6	42	Na_2O	s	−100		17.4
Li	s	0	0	6.7	Na_2O_2	s	−121		22.6
Li_2CO_3	s	−290.5	−270.6	21.6	NaOH	s	−102.1		15.3
LiCl	s	− 97.7	− 92	13.9	Na_2SO_3	s	−260.5	−239.5	34.9
LiF	s	−146	−140	8.6	Na_2SO_4	s	−330.9	−302.8	35.7
LiH	s	− 21.6	− 16.7	5.9	Na_2SiO_3	s	−360	−340	27
Li_3N	s	− 47			Nb	s	0	0	8.6
Mg	s	0	0	7.8	Nb_2O_5	s	− 454		32.8
$MgBr_2$	s	−124	−119		Ni	s	0	0	7.2
$MgCO_3$	s	−265	−245	15.7	NiAl	s	− 34		
$MgCl_2$	s	−153.4	−141.6	21.4	$NiBr_2$	s	− 52		
MgF_2	s	−264	−251	13.7	Ni_3C	s	9		
MgI_2	s	− 86			$NiCl_2$	s	− 75	− 65	25.6
Mg_3N_2	s	−110	− 96		NiO	s	− 58.4	− 51.7	9.2
MgO	s	−144	−136	6.4	NiS	s	22.2		16.1
$Mg(OH)_2$	s	−221	−199	15.1	O_2	g	0	0	49.0
MgS	s	− 83			O	g	59	55	38.5
$MgSO_4$	s	−305	−280	21.9	O_3	g	34	39	57
Mg_2Si	s	− 19			OH	g	10	9	43.9
Mn	α	0	0	7.6	Os	s	0	0	7.8
$MnCO_3$	s	−215	−195	20.5	OsO_4	12)	− 93.5	− 71	29.7
$MnCl_2$	s	−115	−105	28	OsO_4	g	− 80	− 68	65.6
Mn_5N_2	s	− 58	− 47		P	13)	0	0	10.6
MnO	s	− 92	− 87	14.4	P	14)	− 4	− 3	

		ΔH^0	ΔG^0	s^0			ΔH^0	ΔG^0	s^0
P	g	75		39.0	$SrCO_3$	s	−291	−272	23.2
P_2	g	33.8	24.6	52.1	$SrCl_2$	s	−198	−187	
P_4	g	13.1	5.8	66.9	Sr_3N_2	s	− 92		
Pb	s	0	0	15.5	SrO	s	−141	−134	13.0
$PbBr_2$	s	− 66.2	− 62.2	38.6	SrS	s	−108		
$PbCO_3$	s	−167	−150	31.3	$SrSO_4$	s	−345	−319	29.1
$PbCl_2$	s	− 85.8	− 75.0	32.6	Ta	s	0	0	9.9
PbF_2	s	−159	−148		TaC	s	− 38		10.1
PbI_2	s	− 41.8	− 41.5	42.3	TaN·	s	− 58	− 52	
PbO	15)	− 52.4	− 45.2	16.2	Ta_2O_5	s	−499		34
PbO_2	s	− 66	− 52	18,3	Te	s	0	0	11.9
Pb_3O_4	s	−176		50.5	Th	s	0	0	12.8
PbS	s	− 22.5	− 22.1	21.8	ThC_2	s	− 46	− 50	
$PbSO_4$	s	−219.5	−193.9	35.2	$ThCl_4$	s	−285		
Pd	s	0	0	8.9	Th_3N_4	s	−310	−284	
PdO	s	− 20.5			ThO_2	s	−293		
Pt	s	0	0	10.0	Th_2S_3	s	−259		
Rb	s	0	0	18.3	Ti	s	0	0	7.2
Rh	s	0	0	7.6	TiC	s	− 44		5.8
RhO	s	− 22			$TiCl_4$	l	−191.5		60
S	16)	0		7.62	$TiCl_4$	g	−181.6		84
S	17)	0.07	0.015	7.81	TiN	s	− 80	− 73	7.2
S	g	53		40.1	TiO_2	22)	−225.8	−212.5	12.0
S_2	g	31		54.4	Tl	s	0	0	15.4
S_4	g	31		73	Tl_2O	s	− 42	− 33	23.8
S_6	g	27		91	U	s	0	0	12.0
S_8	g	29		110	UC_2	s	− 40		
SF_6	g	−262	−237	69.5	UCl_4	s	−251	−230	47.4
SO_2	g	− 71.0	− 71.8	59.3	UF_6	s	−517	−486	54.5
SO_3	g	− 94.4		61.2	UF_6	g	−505	−485	90.8
Sb	s	0	0	10.9	UN	s	− 80	− 75	
Sb	g	61	51	43.0	UO_2	s	−270	−257	18.6
$SbCl_3$	s	− 91.3	− 77.6	44.5	UO_3	s	−302	−283	23.6
Sb_2O_3	s	−168	−149	29.4	V	s	0	0	7.0
Sb_2O_4	s	−214	−188	30.3	VN	s	− 41	− 34.5	8.9
Sb_2O_5	s	−234	−200	29.9	V_2O_3	s	−295		24
Se	s	0	0	10.1	V_2O_5	s	−380		31
Si	s	0	0	4.5	W	s	0	0	8.0
SiC	s	− 12		3.9	WC	s	− 9		
$SiCl_4$	l	−153.0	−136.9	57.2	W_2N	s	− 17	− 11	
$SiCl_4$	g	−145.7	−136.2	79.2	WO_2	s	−136		
SiF_4	g	−370	−360	68	WO_3	s	−200.8	−182.5	19.9
SiH_4	g	− 15	− 9	49	Zn	s	0	0	9.95
Si_3N_4	s	−179	−155	22	Zn	g	31.2	22.7	38.5
SiO	g	− 22		50.5	$ZnCO_3$	s	−194.2	−175	19.7
SiO_2	18)	−205.4	−192.4	10.0	$ZnCl_2$	s	− 99.4	− 88.3	26
SiO_2	19)	−202.5	−190	11.2	ZnO	s	− 83.2	− 76.0	10.5
Sn	20)	0	0	12.3	ZnS	23)	− 48	− 47	13.8
Sn	21)	− 0.7	0.03	10	ZnS	24)	− 45		
$SnCl_4$	l	−130	−113	62	$ZnSO_4$	s	−234	−208	30.6
SnO	s	− 68	− 61	13.5	Zr	s	0	0	9.2
SnO_2	s	−139	−124	11.6	ZrC	s	− 44		
SnS	s	− 25		18.4	$ZrCl_4$	s	−230	−209	44.5
SnS_2	s	− 40		20.9	ZrN	s	− 87		9.2
Sr	s	0	0	13	ZrO_2	s	−259	−245	12.0

First in the table comes Ag (solid). Here it will be seen that for both ΔH^0 and ΔG^0 the value zero has been inserted. The same is true for the other elements in their stable state. In contrast, the value of the standard entropy is not zero. In this case we are not dealing with the reaction entropy ΔS^0 which for the "reaction" $Ag \rightarrow Ag$ is obviously also zero, but with the molar entropy

$$ s_{Ag}^0 \ (25 \ °C) = \int\limits_0^{298} \frac{c_p dT}{T}. $$

As will be seen below, the great value of the table depends to a large extent on the fact that the data can be so combined that information can be obtained about chemical reactions which have not yet been investigated. If we also had to take into account possible nuclear reactions, i.e. the conversion of one sort of atom into another, we should not be justified in putting ΔH^0 and ΔG^0 equal to zero for all elements. This, in fact, amounts to fixing the zero-point of the enthalpy and the free enthalpy at each element, while, in reality, the various nuclei have widely different energies. Consequently, the table can only be used in ordinary chemistry and metallurgy.

For AgI the table gives the negative values $\Delta H^0 = -14.9$ kcal and $\Delta G^0 = -15.8$ kcal. According to the above, the negative sign of the free enthalpy of formation, ΔG^0, indicates that at 25 °C and 1 atm AgI is stable with respect to solid silver and solid iodine. The negative value of the enthalpy of formation (irreversible heat of formation at constant pressure), ΔH^0, indicates that when 1 mole AgI is formed irreversibly from Ag and I_2 at 25 °C and 1 atm 14.9 kcal heat are released. On the other hand, the reversible heat of formation (the heat which would be absorbed if the reaction took place isothermally and reversibly in a galvanic cell) is given by

$$ Q_{rev} = T\Delta S^0 = \Delta H^0 - \Delta G^0 = 0.9 \ \text{kcal}. $$

The entropy of formation, ΔS^0, is thus $900/298 = 3.0$ cal/deg.mole. This entropy of formation can also be found from the molar entropies of AgI, Ag and $I_2(s)$:

$$ \Delta S^0 = \Sigma \nu s^0 = 27.3 - 10.2 - \frac{27.9}{2} = 3.1 \ \text{cal/deg.mole}. $$

The agreement between the two values of ΔS^0 is not always as satisfactory as in this example. This can be illustrated by calculating these values for AgBr. It should be remembered, however, that both values are obtained as the relatively small difference between large numbers which, furthermore, have generally been found by different investigators.

For AgCN the table gives the positive values $\Delta H^0 = 34$ kcal and $\Delta G^0 = 39$ kcal. The positive sign of ΔG^0 shows that AgCN is metastable with respect to the elements. If the equilibrium were to establish itself, AgCN at 25 °C and 1 atm would decompose spontaneously into silver, graphite and nitrogen. As we have seen in the preceding section, this does not necessarily mean that decomposition would take place at an appreciable rate. We can, however, conclude from the positive sign of ΔG^0 that possible explosivity must be taken into account. A further conclusion is that it is fundamentally impossible to find a catalyst by means of which silver, graphite and nitrogen can be converted into AgCN. For though a catalyst can change the rate of a chemical reaction, it can never change the direction since this would be a contradiction of the second law.

For carbon the table gives information on the modifications graphite and diamond. For graphite we find $\Delta H^0 = 0$ and $\Delta G^0 = 0$. This means that graphite is the stable modification of carbon at 25 °C and 1 atm. In agreement with this ΔG^0 for diamond has a positive value, since it relates to the reaction which produces diamond from graphite. Considered purely from a thermodynamic standpoint, at 25 °C and 1 atm diamond should transform spontaneously into graphite.

The table also shows that Ag_2S, for example, occurs in two modifications, indicated by α and β. The modification which is most stable at 25 °C and 1 atm has the most strongly negative value of ΔG^0.

By means of this table we can also calculate ΔH^0 and ΔG^0 for reactions which are more complicated than the simple combination reactions. We shall illustrate this for a reaction already discussed

$$PbO_2 + 2\ Fe \rightarrow Pb + 2\ FeO. \qquad (3.1.1)$$

Since H and G are thermodynamic functions, ΔH and ΔG are independent of the path followed. These quantities can therefore be calculated for the imaginary path in which PbO_2 is first split into Pb and O_2 and 2 FeO is subsequently formed from 2 Fe and O_2. ΔH^0 (or ΔG^0) for reaction (3.1.1) is thus found by the algebraic summation of the values of ΔH^0 (or ΔG^0) for the reacting substances. This must be done in such a way that for the products the sign is taken from the table, while for the reactants the opposite sign is used. One obtains:

$$\Delta H^0 = -2 \times 64 + 66 = -62 \text{ kcal,}$$
$$\Delta G^0 = -2 \times 58 + 52 = -64 \text{ kcal.}$$

The strongly negative character of ΔG^0 indicates that reaction (3.1.1), thermo-dynamically speaking, has a strong tendency to occur. ΔS^0 can again be found in two ways.

$$\Delta S^0 = \frac{\Delta H^0 - \Delta G^0}{T} = 7 \text{ cal/deg.}$$

$$\Delta S^0 = \Sigma \nu s^0 \qquad = 10 \text{ cal/deg.}$$

The agreement between the two values of ΔS^0 is not worse than is to be expected, since the values of ΔH^0 and ΔG^0 are certainly not more accurately known than to 0.5 kcal, so that $(\Delta H^0 - \Delta G^0)$ may be inaccurate by at least 1 kcal. It should also be noted that we have not taken into account the deviation from stoichiometric composition which always occurs in "FeO" (see Section 4.5).

The free enthalpy of formation of H_2O_2 is seen in the table to be strongly negative, viz. $\Delta G^0 = -27.7$ kcal. This indicates great stability, although it is known that H_2O_2 decomposes spontaneously at 25 °C and 1 atm. The obvious explanation is that ΔG^0 in the table refers only to the stability with respect to the elements, whereas H_2O_2 decomposes into H_2O and O_2:

$$H_2O_2 \to H_2O + \tfrac{1}{2} O_2.$$

For this reaction, according to the table, we have

$$\Delta G^0 = -56.7 + 27.7 = -29.0 \text{ kcal,}$$

so that it can, indeed, take place spontaneously.

The values in the table are only valid for the standard temperature of 25 °C. In the following section we shall see how approximate values of ΔG^0 for other temperatures can be calculated with the help of data from the table. If one wishes to know the value of ΔG^0 for a reaction at a high temperature in which water is involved, one will be dealing with water in the gaseous state, for which the data for $H_2O(l)$ are not applicable. We therefore require to know ΔH^0 and ΔG^0 for imaginary H_2O in the form of a perfect gas at 25 °C and 1 atm. The numbers following $H_2O(g)$ relate to this imaginary gaseous water. They have been obtained by extrapolation. For example, ΔH^0 here gives the enthalpy of formation of gaseous water at 25 °C and 1 atm from 1 mole gaseous H_2 and $\tfrac{1}{2}$ mole gaseous O_2 at the same temperature and pressure.

3.2. Calculation of ΔG^0 for arbitrary temperatures

For arbitrary temperatures, ΔH^0 and ΔG^0 can be calculated from the tabulated values, provided that the specific heats of the reacting substances are known as a function of the temperature. For this purpose we employ the relationships:

$$\Delta H_T = \Delta H_{298} + \int_{298}^{T} \Delta C_p dT, \tag{3.2.1}$$

$$\Delta S_T = \Delta S_{298} + \int_{298}^{T} \frac{\Delta C_p}{T} dT, \tag{3.2.2}$$

where

$$\Delta C_p = \Sigma \nu_i c_{pi}. \tag{3.2.3}$$

The symbol c_p refers to the molar specific heat at constant pressure.

These equations follow directly from the first and second laws when applied to 1 mole of a homogeneous, pure substance. From the first law we have

$$dQ = du + pdv$$

and therefore

$$c_p = \left(\frac{dQ}{dT}\right)_p = \left(\frac{\partial u}{\partial T}\right)_p + p\left(\frac{\partial v}{\partial T}\right)_p$$

or, since $h = u + pv$:

$$c_p = \left(\frac{\partial h}{\partial T}\right)_p. \tag{3.2.4}$$

Integration of (3.2.4) gives:

$$h_T = h_{298} + \int_{298}^{T} c_p dT. \tag{3.2.5}$$

This relationship is valid for 1 mole of each separate substance which takes part in a chemical reaction. By means of equation (2.2.3),

$$\Delta H = \Sigma \nu_i h_i,$$

(3.2.1) can be obtained directly from (3.2.5)

According to the second law

$$ds = \frac{dQ_{\text{rev}}}{T} = \frac{c_p dT}{T}$$

or, after integration:

$$s_T = s_{298} + \int_{298}^{T} \frac{c_p}{T} \, dT .$$ (3.2.6)

By means of the relationship

$$\Delta S = \Sigma \nu_i s_i$$

equation (3.2.2) can be obtained directly from (3.2.6).

If the values of ΔH_T and ΔS_T from equations (3.2.1) and (3.2.2) are substituted in the equation

$$\Delta G_T = \Delta H_T - T\Delta S_T$$ (3.2.7)

we obtain:

$$\Delta G_T = \Delta H_{298} - T\Delta S_{298} + \int_{298}^{T} \Delta C_p dT - T\int_{298}^{T} \frac{\Delta C_p}{T} \, dT.$$ (3.2.8)

Naturally this equation is also valid for reactions which take place under standard conditions. If it is to be applied to a particular reaction, then the molar specific heat must be known as a function of the temperature for each of the substances appearing in the reaction equation. For a great number of elements and compounds these data are to be found in a book by Kubaschewski and Evans [1]) and also in a recent book by Kelley [2]).

In equation (3.2.8) the two terms with the integral sign nearly cancel one another in a great many cases since they have opposite signs and since ΔC_p is usually small. It is therefore often possible to calculate approximate values of ΔG_T by means of the formula

$$\Delta G_T \cong \Delta H_{298} - T\Delta S_{298}.$$ (3.2.9)

Naturally this approximation equation, too, is also applicable to reactions which proceed under standard conditions. It will be applied to many problems in the following chapters. In general it gives better results when the temperature T does not deviate too much from the reference temperature (298 °K).

A better approximation formula, which usually gives reliable results over a wide range of temperature, is obtained when ΔC_p is not wholly neglected

[1]) O. KUBASCHEWSKI and E. LL. EVANS, *Metallurgical Thermochemistry*, Pergamon Press, London, third edition, 1958.
[2]) K. K. KELLEY, Bulletin 584, Bureau of Mines: Contributions to the data on theoretical metallurgy, XIII. High-temperature heat-content, heat-capacity and entropy data for the elements and inorganic compounds, Washington 1960.

but regarded as a constant. Equation (3.2.8) or the more general formula in which 298 °K is replaced by a freely-chosen reference temperature T_r, then becomes

$$\Delta G_T = A + BT \log T + CT, \qquad (3.2.10)$$

where, as can easily be verified:

$$A = \Delta H_{T_r} - T_r \Delta C_p,$$

$$B = -2.3 \, \Delta C_p,$$

$$C = -\Delta S_{T_r} + \Delta C_p \, (1 + \ln T_r).$$

In the above-mentioned book by Kubaschewski and Evans values will be found for the three "constants" A, B and C for a large number of chemical reactions taking place under standard conditions, together with the temperature range to which they are applicable.

3.3. Ellingham diagrams

Equation (3.2.9) shows that by approximation straight lines can be expected when ΔG^0 for chemical reactions is plotted as a function of the temperature. It has, in fact, been found that this is nearly always the case to a very good approximation. Diagrams giving the standard value of the free enthalpy of formation of compounds as a function of the temperature were introduced by Ellingham [1] in 1944. He demonstrated the usefulness of these diagrams, giving those for oxides and sulphides as an example. These are of particularly great importance to metallurgists because many metals occur in their natural state as oxides or sulphides, or as compounds which can easily be converted into oxides. Fig. 2 gives the standard affinity ($-\Delta G^0$) as a function of the temperature for a number of oxidation reactions, which has been deliberately limited for the sake of clarity in the figure. It will be seen that the lines are straight, or virtually so, within the regions where no changes of state occur. These lines can therefore be calculated from the data in Table 1. A more accurate method is to derive them directly from measurements made at higher temperatures.

All the lines in Fig. 2 refer to the reaction with 1 mole of oxygen. The higher a line lies in the figure, the greater the affinity of the element concerned for oxygen. In principle, it is possible to reduce oxides from lower-placed

[1] H. J. T. ELLINGHAM, Trans. Soc. Chem. Ind. 63, 125 (1944).

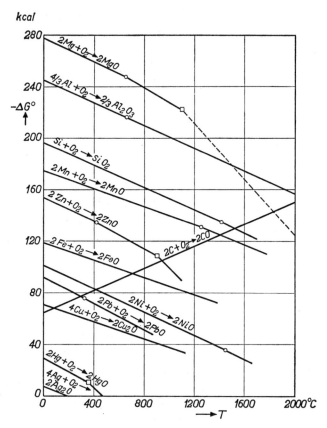

Fig. 2. Standard affinity in kcal of a number of oxidation reactions as a function of temperature. Each circle in the figure corresponds to the melting point, each square to the boiling point of a metal at 1 atm pressure. To the right of the squares the lines refer to the oxidation of gaseous metals (Mg, Zn, Hg) having a pressure of 1 atm.

lines with metals from higher ones. For example, FeO can be reduced with Al, Cu_2O with Fe. The generality of this conclusion is justified because all the reactions relate to 1 mole O_2 so that O_2 vanishes when two reactions are "subtracted" from one another. For example, if (2.1.5) is subtracted from (2.1.6), one obtains the reaction

$$\tfrac{4}{3}\,Al + 2\,FeO \rightarrow \tfrac{2}{3}\,Al_2O_3 + 2\,Fe.$$

For this reaction, according to the figure, $\Delta G^0 \,(= \Delta G_6^0 - \Delta G_5^0)$ really does have a negative value at all temperatures.

Referring all the reactions to the same quantity of oxygen has a second advantage, viz. that when the reactions are carried out electrochemically the same quantity of electricity is transported in each case. For the chosen 1 mole O_2 this quantity is four faradays for each reaction. With the help of equation (2.5.10) thus, the $(-\Delta G^0)$-values which are read off from the figure can be directly converted into values of E^0 by dividing the former by $4 \times 23.06 = 92.2$. Each E^0 value gives the reversible decomposition potential (at a particular temperature) of an oxide in its pure state or in saturated solution in a liquid salt. This is the minimum potential required to decompose an oxide electrolytically into metal and oxygen at 1 atm.

In the extreme lower left-hand corner of the figure is the line for the reaction

$$4 \, Ag + O_2 \rightarrow 2 \, Ag_2O. \tag{3.3.1}$$

It will be noted that below about 200 °C ΔG^0 is weakly negative ($-\Delta G^0$ is weakly positive). The slope of the "straight" line is given, according to (2.2.4), (2.2.5) and (2.3.3), by

$$\frac{\partial(-\Delta G^0)}{\partial T} = \Delta S^0, \tag{3.3.2}$$

i.e. by the entropy change which occurs during the reaction. We have already seen in Section 2.1 that this entropy change is strongly negative, so that the affinity of the reaction decreases with rising temperature. The figure shows that at about 200 °C the standard affinity has decreased to zero. At this temperature, therefore, Ag and Ag_2O are in equilibrium with O_2 at 1 atm; in other words, at this temperature the dissociation pressure of Ag_2O is 1 atm. Above 200 °C Ag_2O decomposes into silver and oxygen even at an oxygen pressure of 1 atm in the surroundings.

If the metal or the oxide undergoes a change of state, the affinity line shows a sudden change of slope. This can be seen, for example, in the second line from bottom left, which relates to the reaction

$$2 \, Hg + O_2 \rightarrow 2 \, HgO. \tag{3.3.3}$$

The change under discussion occurs at 357 °C, the boiling point of mercury. Above this temperature mercury at a pressure of 1 atm is gaseous; its entropy is then much greater than in the liquid state. Consequently, ΔS^0 is more strongly negative above 357 °C than below this temperature and the line slopes more steeply downwards, in accordance with (3.3.2). Melting points and crystallographic transition points, because the corresponding entropy changes are relatively small, produce only slight changes in the direction of the affinity lines. Roughly speaking, for the solid and liquid states of metal and oxide all the lines have approximately the same slope. Obviously this depends on the fact that the entropy change in all these oxidation reactions is determined chiefly by the disappearance of 1 mole O_2.

The exception, which is so important in metallurgy, is the line for the reaction $2\,C + O_2 \rightarrow 2\,CO$, which, for the reasons already discussed in Section 2.1, slopes upwards with rising temperature. Where it cuts another line the standard affinities of the reactions corresponding to the two lines are equal. At temperatures below that at the intersection point the affinity of the metal concerned for oxygen at 1 atm is greater than that of carbon. At temperatures above that of the intersection the reverse is true. This can be illustrated with reference to the point of intersection with the line

$$Si + O_2 \rightarrow SiO_2. \qquad (3.3.4)$$

If we subtract (3.3.4) from (2.1.4), we obtain the reaction

$$SiO_2 + 2\,C \rightarrow Si + 2\,CO. \qquad (3.3.5)$$

The standard affinity of this reaction at the temperature of the intersection has the value zero. Then from (2.4.6) the value of K_p is 1. Since $K_p = p_{CO}^2$, the CO equilibrium pressure of reaction (3.3.5) at the intersection temperature is thus 1 atm. In reality this pressure is already reached at a lower temperature, since there is a fairly great affinity between Si and C, so that the reaction with carbon actually proceeds still further and in accordance with the equation

$$SiO_2 + 3\,C \rightarrow SiC + 2\,CO.$$

This reaction corresponds to greater CO pressures than reaction (3.3.5). Finally we must take into account the formation of gaseous SiO, the result of which is also that the attack of C on SiO_2 becomes noticeable at lower temperatures than would have been expected on the basis of (3.3.5).

In the above we have reasoned as though the reaction between carbon and a metal oxide produces only CO (and no CO_2). As we have shown in Section 2.1, this can be done at high temperatures (above about 900 °C) without introducing large errors.

For diagrams containing a wide selection of curves referring to many oxides [1]), sulphides [2]), carbides [3]), nitrides [4]) and chlorides [5]) the reader is

[1]) F. D. RICHARDSON and J. H. E. JEFFES, J. Iron Steel Inst. **160**, 261 (1948); M. OLETTE and M. F. ANCEY-MORET, Rev. Mét. **60**, 569 (1963).
[2]) F. D. RICHARDSON and J. H. E. JEFFES, J. Iron Steel Inst. **171**, 165 (1952).
[3]) F. D. RICHARDSON, J. Iron Steel Inst. **175**, 33 (1953).
[4]) J. PEARSON and U. J. C. ENDE, J. Iron Steel Inst. **175**, 52 (1953); M. OLETTE and M. F. ANCEY-MORET, Rev. Mét. **60**, 573 (1963).
[5]) H. H. KELLOGG, Trans. AIME **188**, 862 (1950).

referred to the literature on this subject. They have been compiled, after Ellingham, by Richardson and others. Like Table 1, they contain a wealth of thermodynamic data.

3.4. Calculation of equilibrium pressures at 25 °C

By making use of the formula

$$\Delta G^0 = -RT \ln K_p, \tag{3.4.1}$$

many equilibrium pressures at 25 °C, such as the dissociation pressures of compounds of metals and gases, can be calculated directly from the tabulated ΔG^0-values. This will be demonstrated by calculating the dissociation pressure of AgCl. Using the data for $H_2O(l)$ and $H_2O(g)$ we shall also calculate in this section, by way of demonstration, the vapour pressure of water at 25 °C.

(1) For the formation reaction

$$Ag + \tfrac{1}{2} Cl_2 \rightarrow AgCl$$

we have

$$K_p = p_{Cl_2}^{-\frac{1}{2}},$$

$$\Delta G^0 = \tfrac{1}{2} RT \ln p_{Cl_2}.$$

According to the table, the value of ΔG^0 at 25°C is -26.22 kcal, so that the dissociation pressure at that temperature is given by

$$\ln p_{Cl_2} = \frac{2 \, \Delta G^0}{RT} \cong -\frac{2 \times 26220}{2 \times 298} \cong -88.$$

$$p_{Cl_2} \cong 10^{-38} \text{ atm.}$$

The dissociation pressure of AgCl is thus so small at 25 °C that it can be entirely neglected.

If equation (3.4.1) were used in the same manner as above to calculate the dissociation pressure of an endothermal compound like Au_2O_3, a very large value would be found. In this case, however, the use of the equation is not justified because, according to Sections 2.3 and 2.4, it is based on the assumption that the pressures are so small that we may consider the gases as perfect and (even more important) the free enthalpies of the condensed phases as independent of the pressure.

At the very high dissociation pressure which can be calculated for Au_2O_3

at 25 °C by means of equation (3.4.1), about 10^{19} atm, the oxygen would occupy a volume of the same order of magnitude as that of the condensed phases (Au and Au_2O_3). However, we have seen in Sections 2.3 and 2.4 that the derivation of equation (3.4.1) is based on the very assumption that the molar volumes of the condensed phases are negligible with respect to those of the gases.

(2) For the "reaction"

$$H_2O(l) \rightarrow H_2O(g)$$

the table gives for 25 °C and 1 atm

$$\Delta G^0 = -54.63 + 56.69 = 2.06 \text{ kcal.}$$

The vapour pressure of water at 25 °C can now be calculated by means of equation (3.4.1):

$$2.06 = -RT \ln p_{H_2O},$$

$$\ln p_{H_2O} \simeq -\frac{2060}{2 \times 298} \simeq -3.46,$$

$$p_{H_2O} = 0.031 \text{ atm} = 24 \text{ mm Hg.}$$

In reality, of course, the process was carried out in reverse: ΔG^0 for the evaporation reaction in question was calculated from the experimentally determined vapour pressure at 25 °C. The table should be regarded as an adroitly condensed collection of experimental data, from which much can be deduced regarding equilibria which have not been experimentally investigated or which it may even be impossible to investigate.

CHAPTER 4

REACTIONS BETWEEN PURE METALS AND GASES RESULTING IN THE FORMATION OF NEW PHASES

4.1. Approximate calculation of dissociation pressures as functions of the temperature

Many of the equilibrium problems concerning reactions between gases and metals can be brought to an approximate solution by the combined use of the two simple formulae

$$\Delta G_T^0 \cong \Delta H_{298}^0 - T\Delta S_{298}^0, \tag{4.1.1}$$

and

$$\Delta G_T^0 = -RT \ln K_p. \tag{4.1.2}$$

The former is none other than equation (3.2.9) applied to a reaction taking place under standard conditions. The application of the latter formula has been demonstrated in the preceeding section. A combination of the two formulae gives:

$$\log K_p \cong -\frac{\Delta H_{298}^0}{4.575\ T} + \frac{\Delta S_{298}^0}{4.575}. \tag{4.1.3}$$

where ΔH_{298}^0 and ΔS_{298}^0 must be expressed in small calories. We shall now use this formula to calculate the dissociation pressure of Ag_2O as a function of the temperature.

The table gives the following figures for the reaction in which Ag_2O is formed from 2 Ag and $\frac{1}{2}O_2$ at 25 °C:

$$\Delta H^0 = -7.3 \text{ kcal/mole},$$

$$\Delta G^0 = -2.6 \text{ kcal/mole},$$

$$\Delta S^0 = \frac{\Delta H^0 - \Delta G^0}{T} = -15.8 \text{ cal/deg.mole}.$$

Substituting these values in (4.1.3) one obtains the relation:

$$\log p_{O_2} \cong -\frac{3190}{T} + 6.9. \tag{4.1.4}$$

According to this equation the dissociation pressure is zero at absolute zero

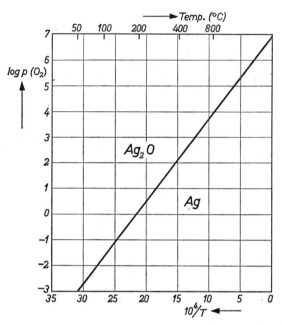

Fig. 3. Dissociation pressure of silver oxide as a function of temperature (log p vs. $1/T$).

temperature and increases continually with rising temperature. At about 190 °C a pressure of 1 atm is reached (cf. Fig. 2). It is seen from the equation that an approximately straight line should be obtained when log p is plotted against $1/T$. The deviations from this linearity, which would be expected from equation (3.2.8), are in many cases so small that it is hardly or not at all possible to verify them experimentally. Figure 3 is a graphical representation of (4.1.4). According to (4.1.3) the slope of the straight line is given by ΔH^0. The intersection of the straight line with the axis $1/T = 0$ ($T = \infty$)is given by ΔS^0.

Ag and Ag₂O can only exist together in equilibrium (co-exist) at temperatures and pressures indicated by points on the straight line. If the oxygen pressure at a certain temperature is smaller than that corresponding to the equilibrium line, the Ag_2O present will decompose into silver and oxygen, if it is higher then the silver present will oxidize to Ag_2O.

In the example discussed we have not only neglected the effect of the specific heats (see Section 3.2), we have not taken into account possible deviations from the stoichiometric composition of Ag_2O nor the fact that silver is able to absorb a quantity of oxygen in solid solution. Even more serious is the fact that melting phenomena have also been left out of consideration. The straight line suffers a change of direction both at the melting point of the metal and at that of the oxide. These changes of direction are given by the heats of fusion (fusion enthalpies) of metal and oxide. For the reaction

$$2 \text{ Me} + \tfrac{1}{2} O_2 \rightarrow Me_2O \qquad (4.1.5)$$

ΔH^0 must be diminished, after the metal (Me) has melted, by twice the molar heat of fusion of Me. After the oxide has melted, however, ΔH^0 must be increased by the heat of fusion of Me_2O. In a graphic representation of the same type as Fig. 3, the slope of the straight line increases when the metal melts, but on the other hand decreases when the oxide melts. This has been shown schematically in Fig. 4 in which point A corresponds to the melting point of the metal and B to the melting point of the oxide. Analogous changes in the slope occur at the boiling points of metal and oxide (cf. also Fig. 2).

In calculating the dissociation pressure of Ag_2O it was assumed that it decomposes into solid or liquid metal and oxygen. It should be borne in mind that the circumstances may be such that *gaseous* metal and oxygen are formed. To demonstrate this we shall look for the dissociation pressure of MgO at 1200 °K. If it is assumed that liquid Mg and O_2 are formed during decomposition according to the equation

$$2 \text{ MgO}(s) \rightleftharpoons 2 \text{ Mg}(l) + O_2, \qquad (4.1.6)$$

then equation (4.1.3) and Table 1 can be used to calculate a reaction constant and thence an oxygen pressure of about 2×10^{-41} atm. Not only the specific heats of the reacting substances, but also the heat of fusion and fusion entropy of magnesium have been neglected in this calculation.

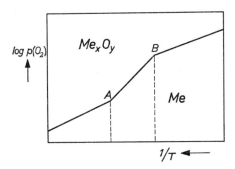

Fig. 4. Schematic representation of the logarithm of the dissociation pressure of an arbitrary metal oxide as a function of the reciprocal of the absolute temperature. Point A corresponds to the melting point of the metal, point B to that of the oxide.

The figure found, 2×10^{-41} atm, would be of the correct order of magnitude if the container in which the reaction takes place really did contain liquid magnesium together with the magnesium oxide. If, however, pure magnesium oxide is used as starting material, the extremely minute dissociation excludes the possibility of the formation of liquid magnesium. At 1200 °K the vapour pressure of liquid Mg has already reached a value of 0.26 atm. Consequently the magnesium will certainly appear during dissociation in the gaseous state. When calculating the dissociation pressure, therefore, we must not start with equation (4.1.6) but with the equation

$$2 \text{ MgO}(s) \rightleftharpoons 2 \text{ Mg}(g) + \text{O}_2, \tag{4.1.7}$$

for which the reaction constant is given by

$$K_p = p_{\text{Mg}}^2 \cdot p_{\text{O}_2}. \tag{4.1.8}$$

This relationship is valid for all Mg-pressures and is thus also true for the case in which liquid Mg in equilibrium with its vapour is present together with the MgO. In that case, according to the above, we have $p_{\text{Mg}} = 0.26$ and $p_{\text{O}_2} = 2 \times 10^{-41}$, so that

$$K_p \cong (0.26)^2 \times 2 \times 10^{-41} \cong 1.3 \times 10^{-42}. \tag{4.1.9}$$

If one starts with pure MgO, two atoms Mg will be formed during dissociation for every molecule O_2:

$$p_{\text{Mg}} = 2 p_{\text{O}_2}.$$

If we substitute this in (4.1.8), we obtain, making use of (4.1.9):

$$K_p = 4 p_{\text{O}_2}^3 \cong 1.3 \times 10^{-42},$$

$$p_{\text{O}_2} \cong 6.5 \times 10^{-15},$$

$$p_{\text{Mg}} \cong 13 \times 10^{-15}.$$

The true oxygen pressure is seen from this result to be about 10^{26} times as large as was calculated on the basis of equation (4.1.6).

4.2. Reaction of metals with wet hydrogen

Pure chromium is heated in hydrogen. How dry must the hydrogen be at various temperatures in order to avoid oxidation of the chromium?

Oxidation, when it occurs, takes place according to the reaction

$$2 \, Cr + 3 \, H_2O \rightleftharpoons Cr_2O_3 + 3 \, H_2. \qquad (4.2.1)$$

Table 1 gives the following figures, valid when the reaction takes place at 25 °C and 1 atm (with gaseous water!):

$$\Delta H^0 = -269.7 + 3 \times 57.8 = -96.3 \text{ kcal,}$$

$$\Delta G^0 = -250.2 + 3 \times 54.63 = -86.3 \text{ kcal,}$$

$$\Delta S^0 = \frac{\Delta H^0 - \Delta G^0}{T} = -33.6 \text{ cal/deg.}$$

By means of the relationship $\Delta S^0 = \Sigma \nu_i s_i$ virtually the same standard value of the reaction entropy is found. If the above values are inserted in equation (4.1.3), we obtain:

$$\log \left(\frac{p_{H_2}}{p_{H_2O}} \right)^3 \simeq \frac{21050}{T} - 7.35$$

or

$$\log \frac{p_{H_2O}}{p_{H_2}} \simeq -\frac{7017}{T} + 2.45. \qquad (4.2.2)$$

Table 2 gives the equilibrium water vapour content of hydrogen calculated for various temperatures from the above formula.

If the chromium is to remain bright when heated in hydrogen, the gas at 1000 °K may not contain very much more than one hundred-thousandth part (0.001 %) of water vapour. At higher temperatures the hydrogen may be much wetter without causing oxidation. On the other hand, as the temperature drops the hydrogen is required to be much drier. A hydrogen atmosphere with an H_2O-content of e.g. 10^{-10} % is not attainable in practice. Chromium bright-annealed at high temperatures will thus always oxidize superficially when it cools down with the furnace in purified hydrogen. Therefore, after it has been bright-annealed in hydrogen it must be slid as quickly as possible to a cold part of the furnace tube.

TABLE 2

WATER VAPOUR CONTENTS OF HYDROGEN AT WHICH CHROMIUM BEGINS TO OXIDIZE
(ROUGH CALCULATION)

Temp. °K	Temp. °C	$\dfrac{p_{H_2O}}{p_{H_2}}$	H_2O vol. %
500	227	3×10^{-12}	3×10^{-10}
1000	727	3×10^{-5}	0.003
1500	1227	6×10^{-3}	0.6
2000	1727	9×10^{-2}	8

It is interesting to repeat the calculations, carried out above for a reactive metal such as chromium, for a more noble metal such as copper. For the reaction

$$2 \, Cu + H_2O \rightleftharpoons Cu_2O + H_2 \qquad (4.2.3)$$

the following values are valid, according to Table 1, at 25 °C and 1 atm:

$$\Delta H^0 = -40 + 57.8 = 17.8 \text{ kcal},$$

$$\Delta G^0 = -35.0 + 54.6 = 19.6 \text{ kcal},$$

$$\Delta S^0 = \frac{\Delta H^0 - \Delta G^0}{T} = -6.04 \text{ cal/deg.}$$

Substitution in equation (4.1.3) gives:

$$\log \frac{p_{H_2}}{p_{H_2O}} \simeq -\frac{3890}{T} - 1.32. \qquad (4.2.4)$$

The equilibrium hydrogen contents calculated from this equation for two temperatures are given in Table 3.

TABLE 3

HYDROGEN CONTENT OF $(H_2 + H_2O)$ MIXTURES IN EQUILIBRIUM WITH $(Cu + Cu_2O)$
(ROUGH CALCULATION)

Temp. °K	Temp. °C	$\dfrac{p_{H_2}}{p_{H_2O}}$	H_2 vol. %
500	227	8×10^{-10}	8×10^{-8}
1000	727	6×10^{-6}	6×10^{-4}

Whereas the mixture $(H_2 + H_2O)$ which is in equilibrium with $(Cr + Cr_2O_3)$ is seen from Table 2 to consist almost entirely of hydrogen, for $(Cu + Cu_2O)$ the situation is reversed: the equilibrium mixture of gases is shown by the last table to consist almost entirely of steam. Steam with only 0.001 % H_2 has a reducing action on Cu_2O at 1000 °K. The question arises whether steam at, say, 1000 °K does not already contain enough hydrogen from thermal dissociation to have a reducing action on Cu_2O. From the data in Table 1 it can be calculated that this is not the case.

Perhaps we should repeat that the calculations carried out in this and the preceding section are only approximations. In the first place because it has been assumed that ΔC_p (see Section 3.2) may be neglected, in the second place because the liquid and solid substances concerned were regarded as pure, i.e. as being unable to dissolve even the smallest quantity of the substances with which they are in contact in the equilibria discussed. These approximations, however, introduce no grave errors in the cases concerned.

THE DEPENDENCE OF THE REACTION CONSTANT ON THE TEMPERATURE

It has been calculated above that the humidity of hydrogen in equilibrium with chromium plus chromium oxide increases with rising temperature, in equilibrium with copper plus copper oxide, on the other hand, it decreases. The reason for this different dependence on temperature lies in the fact that reaction (4.2.1) is exothermic (ΔH^0 negative), while reaction (4.2.3) is endothermic. It is known from equation (2.4.6) that

$$\ln K_p = -\frac{\Delta G^0}{RT} = -\frac{\Delta H^0}{RT} + \frac{\Delta S^0}{R}, \qquad (4.2.5)$$

so that $\ln K$, and with it K, decreases with rising temperature if ΔH^0 is negative, but increases if ΔH^0 is positive.

This is not only true in the case where ΔH^0 and ΔS^0 may be regarded as virtually constant, but also in the general case where they are functions of the temperature. By differentiating with respect to the temperature at constant pressure we obtain:

$$\frac{d \ln K_p}{dT} = \frac{\Delta G^0}{RT^2} - \frac{1}{RT}\left(\frac{\partial \Delta G^0}{\partial T}\right)_p.$$

Employing equation (3.3.2):

$$\frac{d \ln K_p}{dT} = \frac{\Delta G^0}{RT^2} + \frac{T\Delta S^0}{RT^2}$$

or

$$\frac{d \ln K_p}{dT} = \frac{\Delta H^0}{RT^2}. \qquad (4.2.6)$$

This is the well-known rule of Van 't Hoff for the shifting of a chemical equilibrium which accompanies a change of temperature. It shows that the direction in which the shift occurs indeed depends only on the sign of ΔH^0.

Equation (4.2.6) can, of course, also be derived by differentiating each of the terms $-\Delta H^0/RT$ and $\Delta S^0/R$ in (4.2.5) separately with respect to the temperature and showing that the effects of the specific heats exactly compensate each other.

As already discussed in Section 4.1 for a special case, a straight line is always obtained when log K is plotted against $1/T$ over a not-too-large range of temperature. In that range, therefore, ΔH^0 may be regarded as a constant, which is given by the slope of the line (multiplied by -4.575). The point of intersection of the straight line with the axis $1/T = 0$ gives (with the exception of the factor 4.575) the magnitude of ΔS^0.

4.3. Steam-embrittlement of copper

The easy reducibility of Cu_2O, discussed in Section 4.2, enables us to understand the so-called steam-embrittlement of copper. This term describes the brittleness which occurs in copper as a result of heating in hydrogen or a mixture of gases containing hydrogen [1]). The phenomenon is only encountered in copper which contains oxygen. We shall therefore briefly discuss the copper-oxygen system.

Fig. 5 shows part of the phase diagram of this system. The solubility of oxygen in liquid copper is fairly large, that in solid copper, on the other hand, is very small. The maximum solubility of oxygen in solid copper occurs at the eutectic temperature of about 1065 °C; according to the most reliable data it does not amount to more than 0.01 % by weight at most. It decreases with decreasing temperature and at 600 °C it has already dropped to less than 0.002 %. If the copper contains impurities with a greater affinity for oxygen (e.g. iron), these will be converted into oxides by internal oxidation when oxygen is dissolved. One can then measure apparent solubilities which may be considerably larger than the real solubility. This is the explanation for the very contradictory data on the solubility as given in the literature. The only solid oxides which occur in the system are Cu_2O and CuO. Contrary to the older data, Cu_2O does *not* decompose below about 375 °C into Cu and CuO. This can be seen from Table 1 according to which ΔG^0 for the reaction

$$Cu_2O \nrightarrow Cu + CuO$$

has a positive value at 25 °C. When copper containing more than 0.01 % O solidifies, one generally obtains a polycrystalline material along the crystal boundaries of which perceptible quantities of the eutectic mixture (Cu + Cu_2O) are present.

Like oxygen hydrogen has a small solubility in copper. If copper is heated in this gas, a portion diffuses into the interior and reduces the Cu_2O which it encounters. According to the available data on atomic weights and

---

[1]) See e.g. C. E. RANSLEY, J. Inst. Metals **65**, 147 (1939).

PLATE 1

Fig. 6. Cracks in the surface of oxygen-bearing copper after heating in hydrogen at 850 °C. Before the treatment in hydrogen the surface was polished. (Magnification 100×). From H. T. SCHAAP, Metalen **12**, 204 (1957).

Fig. 7a. Blisters on a strip of copper containing oxygen after heat treatment in hydrogen at 850 °C (Magnification 2×). From H. T. SCHAAP, Metalen **12**, 204 (1957).

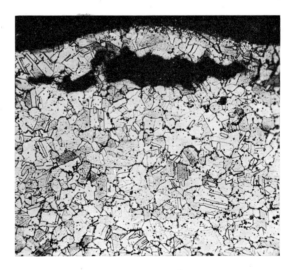

Fig. 7b. Cross-section of one of the blisters in Fig. 7a (Magnification 50×).

PLATE 2

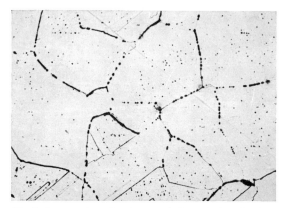

Fig. 8. Pores and cracks along grain boundaries after heat treatment of copper containing oxygen in hydrogen at 850 °C (Magnification 100×). From H. T. SCHAAP, Metalen **12**, 204 (1957).

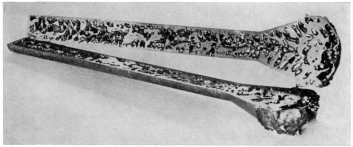

Fig. 33. Sawn-through bar of pure iron, obtained by melting and casting in hydrogen at 1 atm. Owing to the sharp drop in solubility upon solidification, the metal contains numerous gas-filled cavities.

Fig. 35. Blisters formed during pickling of steel. In the pickling process, atomic hydrogen is produced which diffuses inwards and forms molecular hydrogen at inclusions in the steel. The blisters appear where these inclusions are immediately below the surface. Magnification 10×. (Fast and Van Ooijen.)

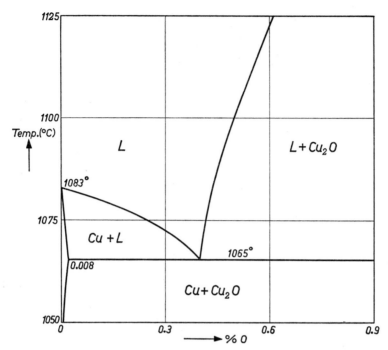

Fig. 5. Part of the copper-oxygen phase diagram. The oxygen content in wt% is plotted along the horizontal axis. (A. BUTTS, *Copper*, American Chem. Soc. Monograph Series, Reinhold Publ. Corp., New York, 1954).

specific gravities, the reduction of a volume of 23.8 cm³ Cu₂O (1 mole Cu₂O) produces a volume of 14.3 cm³ Cu (2 mole Cu). Cavities are thus formed in which water vapour collects. The pressure of this water vapour can be calculated very roughly by supposing that it behaves as a perfect gas, that no water vapour escapes by diffusion and that the copper can withstand the water vapour pressure without deforming or breaking. In this case a volume of 23.8 — 14.3 = 9.5 cm³ is available for each mole of H_2O which is formed. From these figures with the help of the law of Boyle-Gay Lussac we can immediately calculate an H_2O pressure in the cavities of about 10^4 atm at 850 °C. Furthermore, in the equilibrium state there will be an H_2 pressure in the cavities equal to that outside the metal. If this is 1 atm, the value of p_{H_2}/p_{H_2O} in the cavities will be about 10^{-4}.

The calculations above would be meaningless if a mixture of H_2 and H_2O in this pressure ratio were unable to reduce Cu₂O. Equation (4.2.4) shows, however, that a pressure ratio of 10^{-4} is still amply sufficient to reduce Cu₂O at 850 °C. Thermodynamically speaking, therefore, it is possible for the high

H_2O pressure just calculated to occur. The condition that no water vapour escapes by diffusion is fulfilled reasonably well, since the rate of transport of water vapour in copper is many times smaller than that of hydrogen [1]. The condition that the copper must be capable of withstanding the high pressure in the cavities is certainly not fulfilled at high temperatures. Many crystals where Cu_2O is found along the boundaries will be torn apart long before a pressure of 10^4 atm is reached. This is demonstrated by Figure 6 which shows the surface of a piece of copper containing oxygen which has been polished before annealing in hydrogen (see Plate 1 opposite page 54, which also shows Figs. 7a and 7b).

When copper containing oxygen is heated in hydrogen, not only do cracks appear along the grain boundaries but it is also frequently observed that blisters arise on the surface. These are caused by reduction of Cu_2O which is present not far below the surface. Fig. 7a shows these blisters on a strip of copper, while Fig. 7b gives a cross-section through one of the blisters. Examination of steam-embrittled copper under a microscope often reveals not only cracks, but also rows of pores along the grain boundaries. These spring from Cu_2O which was present in the form of globules along the boundaries. Figure 8 (opposite p. 55) which, like the three preceding figures, has been derived from an article by Schaap [2], illustrates this phenomenon. Cavities in the interior of the crystals are also not exceptional in steam-embrittled copper, particularly in coarse-grained copper.

If copper containing oxygen is heated for a long time in hydrogen at a relatively low temperature (e.g. 250 °C), the metal is subsequently not brittle, but the embrittlement will be latent in it. As was the case at high temperatures the Cu_2O present is reduced at 250 °C by the penetrating hydrogen, but the vapour pressure of the water formed (critical temperature 374 °C) is not high enough to produce cracks in the metal. If this metal is then heated for a short time at a high temperature (e.g. 700 °C) in vacuum or nitrogen, all the water will be converted into the gaseous form and the high H_2O pressure causes cracks along the grain boundaries, which implies that the copper becomes brittle [3].

The difficulties described, which occur when copper containing oxygen is heated in hydrogen, can be avoided by choosing CO instead of H_2 as the reducing gas. The former gas is insoluble in copper and therefore unable to reduce Cu_2O inclusions directly. Nevertheless, if oxygen-bearing copper

---

[1] J. H. DE BOER and J. D. FAST, Rec. trav. chim. Pays-Bas **54**, 970 (1935).

[2] H. T. SCHAAP, Metalen **12**, 204 (1957); in Dutch.

[3] S. HARPER, V. A. CALLCUT, D. W. TOWNSEND and R. EBORALL, J. Inst. Metals **90**, 414 and 423 (1962).

is heated long enough in CO it is possible to remove the oxygen completely because the dissolved oxygen reacts with CO at the surface, forming CO_2. This causes a diffusion current of oxygen atoms (or ions) from the Cu_2O precipitate to the outer surface until, eventually, all the Cu_2O has disappeared.

4.4. Exact calculation of chemical equilibrium

As was shown in Section 3.2, the position of an equilibrium at high temperatures can be calculated, more accurately than we have already done, by making use of the specific heats of the reaction partners. We shall demonstrate this with the equilibrium

$$2 \, Cr + 3 \, H_2O \rightleftharpoons Cr_2O_3 + 3 \, H_2, \qquad (4.4.1)$$

the location of which was roughly calculated in Section 4.2 for a few temperatures. We shall first recalculate this equilibrium, no longer neglecting ΔC_p, the change in thermal capacity caused by the process of the reaction, but regarding it as a constant. After that we shall calculate the equilibrium as accurately as possible, i.e. as accurately as the available data on specific heats permit.

(a) ΔC_p is regarded as a constant

According to Section 3.2, the free enthalpy of reaction is given in this case by

$$\Delta G_T = A + BT \log T + CT, \qquad (4.4.2)$$

in which A, B and C are constants. The book by Kubaschewski and Evans [1]), which has already been referred to before, does not give the values of these constants for equation (4.4.1), but does give them for the reactions in which H_2O and Cr_2O_3 decompose into their constituents. The required values of A, B and C can be derived from these without any difficulty:

$3 \, H_2O \rightleftharpoons 3 \, H_2 + \frac{3}{2} O_2; \quad A = 171\,750, \qquad B = -13.44, \qquad C = 6.63$

$Cr_2O_3 \rightleftharpoons 2 \, Cr + \frac{3}{2} O_2; \quad A = 267\,750, \qquad\qquad\qquad\qquad C = -62.1$

Subtracting, we obtain

$2 \, Cr + 3 \, H_2O \rightleftharpoons Cr_2O_3 + 3 \, H_2; \quad A = -96\,000, \; B = -13.44, \; C = 68.73$

---

[1]) O. KUBASCHEWSKI and E. LL. EVANS, *Metallurgical Thermochemistry*, Pergamon Press, London, third edition, 1958.

Substituting these values in equation (4.4.2) and employing the well-known equation

$$\Delta G_T = - RT \ln K_p,$$

one finds the values given in Table 9, for p_{H_2O}/p_{H_2}.

(b) ΔC_p is dependent on the temperature

For an exact calculation of the equilibrium in question, account must be taken of the temperature dependence of the specific heats of the reaction partners. We can conveniently employ the recent compilation by Kelley [1] of the specific heats of the elements and inorganic compounds. For the substances occurring in equation (4.4.1) this book gives the following:

$$c_p(\text{Cr},s) \quad = \quad 4.16 + 3.62 \times 10^{-3}T + 0.30 \times 10^5 T^{-2}, \qquad (4.4.3)$$

$$c_p(\text{Cr}_2\text{O}_3,s) = 28.53 + 2.20 \times 10^{-3}T - 3.74 \times 10^5 T^{-2}, \qquad (4.4.4)$$

$$c_p(\text{H}_2,g) \quad = \quad 6.52 + 0.78 \times 10^{-3}T + 0.12 \times 10^5 T^{-2}, \qquad (4.4.5)$$

$$c_p(\text{H}_2\text{O},g) \quad = \quad 7.30 + 2.46 \times 10^{-3}T. \qquad (4.4.6)$$

These equations give the heat-capacity per mole in cal/deg. Using these equations and also equation (3.2.8) and Table 1, it is not difficult to cal-

TABLE 4

ENTHALPY AND ENTROPY OF Cr(s,l)

$T(°K)$	$h^0{}_T - h^0{}_{298}$ cal/mole	$s^0{}_T - s^0{}_{298}$ cal/deg.mole	$T(°K)$	$h^0{}_T - h^0{}_{298}$ cal/mole	$s^0{}_T - s^0{}_{298}$ cal/deg.mole
400	595	1.72	1700	10930	12.51
500	1220	3.11	1800	11980	13.11
600	1870	4.29	1900	13080	13.71
700	2530	5.31	2000	14220	14.29
800	3210	6.22	2100	15410	14.87
900	3910	7.03	2176(s)	16340	15.31
1000	4640	7.81	2176(l)	21340	17.61
1100	5410	8.54	2200	21570	17.71
1200	6230	9.26	2400	23450	18.53
1300	7100	9.95	2600	25330	19.28
1400	8010	10.63	2800	27210	19.98
1500	8950	11.28	3000	29090	20.63
1600	9920	11.90			

[1] K. K. KELLEY, Bulletin 584, Bureau of Mines: Contributions to the data on theoretical metallurgy, XIII. High-temperature heat-content, heat-capacity and entropy data for the elements and inorganic compounds, Washington 1960.

TABLE 5

ENTHALPY AND ENTROPY OF $Cr_2O_3(s)$

$T(°K)$	$h^0{}_T - h^0{}_{298}$ cal/mole	$s^0{}_T - s^0{}_{298}$ cal/deg.mole	$T(°K)$	$h^0{}_T - h^0{}_{298}$ cal/mole	$s^0{}_T - s^0{}_{298}$ cal/deg.mole
400	2740	7.94	1300	29550	42.60
500	5540	14.19	1400	32670	44.91
600	8380	19.36	1500	35790	47.07
700	11280	23.82	1600	38920	49.08
800	14230	27.76	1700	42050	50.98
900	17210	31.27	1800	45180	52.77
1000	20240	34.46	1900	48320	54.47
1100	23320	37.40	2000	51460	56.08
1200	26430	40.11			

TABLE 6

ENTHALPY AND ENTROPY OF $H_2(g)$

$T(°K)$	$h^0{}_T - h^0{}_{298}$ cal/mole	$s^0{}_T - s^0{}_{298}$ cal/deg.mole	$T(°K)$	$h^0{}_T - h^0{}_{298}$ cal/mole	$s^0{}_T - s^0{}_{298}$ cal/deg.mole
400	705	2.04	2000	12650	13.79
500	1405	3.60	2100	13470	14.19
600	2105	4.87	2200	14300	14.58
700	2810	5.96	2300	15135	14.95
800	3515	6.90	2400	15975	15.31
900	4225	7.74	2500	16825	15.66
1000	4940	8.49	2750	18980	16.48
1100	5670	9.18	3000	21160	17.24
1200	6405	9.82	3250	23375	17.94
1300	7150	10.42	3500	25610	18.61
1400	7905	10.98	3750	27870	19.23
1500	8670	11.51	4000	30150	19.82
1600	9445	12.01	4250	32445	20.38
1700	10235	12.49	4500	34755	20.90
1800	11030	12.94	4750	37085	21.41
1900	11835	13.38	5000	39425	21.89

culate exactly the position of the equilibrium under discussion for several temperatures. One can save oneself the trouble of the integrations required by equation (3.2.8) by making use of the tables of standard values of the molar enthalpies and entropies with reference to 298 °K, which are also to be found in Kelley's book. To emphasize the great importance of these tables, we have reprinted those for Cr(s,l), $Cr_2O_3(s)$, $H_2(g)$ and $H_2O(g)$ in their entirety (Tables 4, 5, 6 and 7).

It is easily seen that the reaction enthalpy of (4.4.1) at an arbitrary temperature T is given by:

TABLE 7

ENTHALPY AND ENTROPY OF $H_2O(g)$

$T(°K)$	$h^0_T - h^0_{298}$ cal/mole	$s^0_T - s^0_{298}$ cal/deg.mole	$T(°K)$	$h^0_T - h^0_{298}$ cal/mole	$s^0_T - s^0_{298}$ cal/deg.mole
400	825	2.38	2000	17370	18.13
500	1655	4.23	2100	18600	18.73
600	2510	5.79	2200	19845	19.31
700	3390	7.14	2300	21100	19.87
800	4300	8.36	2400	22370	20.41
900	5240	9.47	2500	23650	20.93
1000	6210	10.49	2750	26895	22.17
1100	7210	11.44	3000	30200	23.32
1200	8240	12.34	3250	33545	24.38
1300	9295	13.18	3500	36930	25.38
1400	10385	13.99	3750	40350	26.33
1500	11495	14.76	4000	43805	27.22
1600	12630	15.49	4250	47275	28.06
1700	13785	16.19	4500	50770	28.86
1800	14965	16.86	4750	54290	29.62
1900	16160	17.51	5000	57825	30.34

$$\Delta H^0_T = \Delta H^0_{298} + (h^0_T - h^0_{298})_{Cr_2O_3} + 3(h^0_T - h^0_{298})_{H_2} - 2(h^0_T - h^0_{298})_{Cr}$$
$$- 3 (h^0_T - h^0_{298})_{H_2O}. \qquad (4.4.7)$$

The quantity ΔH^0_{298} which appears above, was shown in Section 4.2 to have the value -96300 cal. For the reaction entropy an equation analogous to (4.4.7) can be used. In this equation $\Delta S^0_{298} = -33.6$ cal/deg. The values of ΔH^0 and ΔS^0, calculated in this way for the temperatures in Table 2, are to be found in Table 8. This table also contains the values of the equilibrium quotient p_{H_2O}/p_{H_2} calculated with the help of the well-known equation

$$\log K_p(T) = -\frac{\Delta H^0_T}{4.575\ T} + \frac{\Delta S^0_T}{4.575}.$$

TABLE 8

DATA RELATING TO EQUILIBRIUM (4.4.1) AND OBTAINED BY "EXACT" CALCULATION

$T(°K)$	ΔH^0 cal	ΔS^0 cal/deg	$\log K_p$	p_{H_2O}/p_{H_2}
500	− 93950	− 27.52	35.06	2.04×10^{-12}
1000	− 89150	− 20.76	14.95	1.05×10^{-5}
1500	− 86885	− 18.84	8.54	1.41×10^{-3}
2000	− 87440	− 19.12	5.38	1.62×10^{-2}

Finally, in Table 9 we find the collected equilibrium values of p_{H_2O}/p_{H_2} for equation (4.4.1), which have been calculated by the three methods discussed above. It will be seen that the rough method of calculation produces values which deviate more from the real values when the temperature differs more from 298 °K.

TABLE 9

VALUES OF p_{H_2O}/p_{H_2} CALCULATED IN THREE DIFFERENT WAYS FOR THE EQUILIBRIUM
$2\ Cr + 3\ H_2O \rightleftharpoons Cr_2O_3 + 3\ H_2$

$T(°K)$	p_{H_2O}/p_{H_2} approx. ($\Delta C_p = 0$)	p_{H_2O}/p_{H_2} ΔC_p const.	p_{H_2O}/p_{H_2} "exact"
500	3×10^{-12}	2.3×10^{-12}	2.04×10^{-12}
1000	3×10^{-5}	1.2×10^{-5}	1.05×10^{-5}
1500	6×10^{-3}	1.7×10^{-3}	1.41×10^{-3}
2000	9×10^{-2}	1.9×10^{-2}	1.62×10^{-2}

4.5. The iron-oxygen system

Figure 9 gives part of the phase diagram of the important iron-oxygen system, according to Darken and Gurry [1]). Some modifications have been introduced by the author with reference to later investigations by Engell [2]). Oxygen exhibits a distinct solubility in liquid iron, although this is not visible in Fig. 9 because this only shows the phase-relationships in the concentration range between 49 and 61 atom% oxygen. On the other hand, oxygen is virtually insoluble in the modifications of *solid* iron. According to the most reliable data, the solubility at high temperatures, in both body-centred cubic (α and δ) and face-centred cubic (γ) iron, is not more than about one hundredth weight percent at most. For temperatures below 600 °C it can be regarded as practically equal to zero (see section 7.13, small print).

Three solid oxide phases occur in the system: Fe_xO (wüstite), Fe_3O_4 (magnetite) and Fe_2O_3 (hematite). Whereas Fe_3O_4 and Fe_2O_3 only exist within a small region of concentrations, Fe_xO exhibits a wide homogeneity range. A most remarkable fact is that FeO of stoichiometric composition is unstable [3]). It can be seen from Fig. 9 that at 1000 °C the region of exist-

[1]) L. S. DARKEN and R. W. GURRY, J. Amer. Chem. Soc. 68, 798 (1946).
[2]) H. J. ENGELL, Arch. Eisenhüttenwes. 28, 109 (1957).
[3]) See e.g. P. K. FOSTER and A. J. E. WELCH, Trans. Faraday Soc. 52, 1626 (1956). This article also contains a critical survey of the older literature.

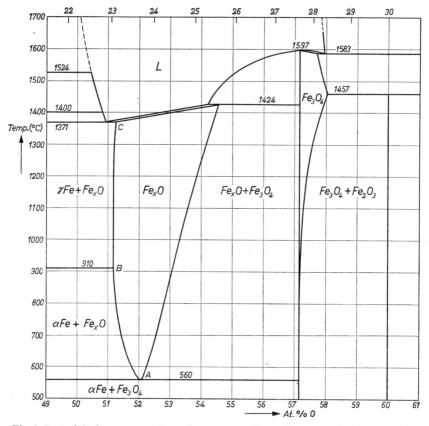

Fig. 9. Part of the iron-oxygen phase diagram according to DARKEN and GURRY, J. Amer. Chem. Soc. **68**, 798 (1946), but slightly modified by the author on the basis of measurements by ENGELL, Arch. Eisenhüttenwes, **28**, 109 (1957). The lower horizontal axis gives the oxygen content in atom%, the upper one in wt%.

ence of the wüstite phase extends from 1.05 atoms O per atom Fe, when wüstite is in equilibrium with Fe, to 1.14 atoms O per atom Fe, when it is in equilibrium with Fe_3O_4. Furthermore, X-ray examination has shown that wüstite should not be regarded as FeO with an excess of oxygen, but rather as FeO with a deficiency of iron. It has the same crystal structure as NaCl, but the oxygen sites in the lattice are fully occupied while the iron sites are not. In equilibrium with metallic iron the number of unoccupied sites above 800 °C is virtually independent of the temperature; it corresponds to the composition $Fe_{0.95}O$. In other words: above 800 °C wüstite may be considered as a phase of almost constant composition if it is in equilibrium with iron.

Which equilibrium states will occur if pure iron at constant temperature is brought into contact with oxygen at gradually increasing pressure? To answer this question we suppose that a piece of pure iron is situated in an absolute vacuum at a temperature of 1000 °C. If the oxygen pressure is gradually allowed to rise from zero, the following changes will take place (see schematic diagram in Fig. 10). At first a small quantity of oxygen will be absorbed in solid solution. The quantity dissolved will increase with the pressure (curve Fe). As will be seen in Chapter 7, this kind of solubility curve of a diatomic gas in a metal is parabolic in form, since the gas is monatomic inside the metal. At point A, corresponding to the pressure p_1 and the oxygen content O_1, the limit of solubility is reached. According to the solubility measurements mentioned above, point A lies virtually on the vertical axis. At pressure p_1, the lowest oxide phase, with the composition $Fe_{0.95}O$, begins to form in co-existence with the metallic phase. Besides the gaseous phase, the system now comprises two solid phases and thus, according to the phase rule, at a particular temperature it is only in equilibrium at one particular gas pressure. The pressure will thus remain constant until all the metal has been converted into $Fe_{0.95}O$. To raise the oxygen content still further, the pressure must increase (curve Fe_xO) until finally the upper limit O_3 (Fig. 10) of the existence region of Fe_xO is reached at a pressure p_2. The rising oxygen content along this curve corresponds to an increasing fraction of unoccupied

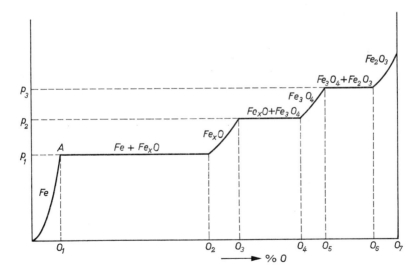

Fig. 10. Schematic representation of the equilibrium pressure of oxygen over iron at 1000 °C as a function of the quantity of oxygen taken up by this metal.

iron sites in Fe_xO, i.e. with a decrease in x. At pressure p_2, Fe_3O_4 begins to form in co-existence with Fe_xO, and so on.

Below 560 °C, wüstite is unstable in the equilibrium state, so that Fe_3O_4 forms immediately on the iron as it oxidizes. The fact that wüstite is unstable at room temperature, can be deduced by means of Table 1 directly from the fact that ΔG^0 has a negative value for the reaction

$$4\ Fe_{0.95}O \longrightarrow Fe_3O_4 + 0.8\ Fe.$$

In each of the regions with two co-existent solid phases (horizontal sections of the curve in Fig. 10) the equilibrium pressure is identical with the dissociation pressure of the higher oxide, or, more precisely, with the dissociation pressure associated with the oxygen-poor limit of this higher oxide. As a measure of these dissociation pressures, Fig. 11 shows $RT \ln p_{O_2}$ as a function of the temperature. The four straight lines in the figure relate to the reactions

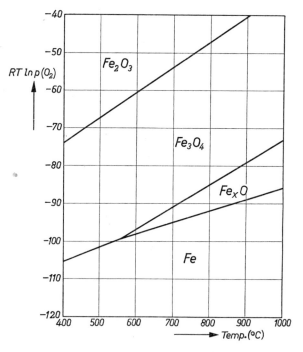

Fig. 11. Dissociation pressures of the oxides of iron as functions of the temperature ($RT \ln p$ in kcal vs. temperature in °C).

$$6\ Fe_2O_3 \rightleftharpoons 4\ Fe_3O_4 + O_2, \tag{4.5.1}$$

$$\tfrac{1}{2}\ Fe_3O_4 \rightleftharpoons \tfrac{3}{2}\ Fe\quad + O_2, \tag{4.5.2}$$

$$2\ Fe_3O_4 \rightleftharpoons 6\ FeO\quad + O_2, \tag{4.5.3}$$

$$2\ FeO\quad \rightleftharpoons 2\ Fe\quad + O_2. \tag{4.5.4}$$

The last two reaction formulae are not absolutely correct, since Fe_xO has been written FeO for the sake of simplicity. The equilibrium

$$2\ Fe_2O_3 \rightleftharpoons 4\ Fe + 3\ O_2$$

does not exist. The phase diagram and Fig. 11 both show that Fe_2O_3 and Fe can never be in equilibrium with one another. The same is true for Fe_2O_3 and Fe_xO. At point A in the phase diagram there is equilibrium between four phases: Fe, Fe_xO, Fe_3O_4 and the gas phase. At the temperature corresponding to A (560 °C) the equilibrium pressure of O_2 is therefore the same for the three reactions (4.5.2), (4.5.3) and (4.5.4). This appears in Fig. 11 as a common point of the three pressure lines (at 560 °C). If one only has the data in Table 1 at his disposal, then this result can not be obtained by calculation. This can only be done by means of more subtle calculations employing equation (3.2.8).

In the above, we have discussed equilibrium states. But if iron at, say, 1000 °C is brought into contact with oxygen at such a pressure that all the iron must eventually be converted into Fe_2O_3, then an oxide film will be formed at first on the surface of the metal, consisting of Fe_xO, Fe_3O_4 and Fe_2O_3 (see Fig. 12). This can be explained by assuming that even in non-equilibrium states, equilibrium will always be established at the interfaces at high temperatures. And if that is the case, it follows from the above that Fe_2O_3 cannot be in contact with Fe or Fe_xO, but only with Fe_3O_4. Also the latter oxide cannot be in direct contact with Fe at 1000 °C. The relative quantities of the three oxides formed are a function of time and temperature.

In the second volume of this work it will be seen that the oxidation proceeds mainly by the diffusion of ferro-ions through the oxide to the outer surface where they form new oxide in combination with oxygen. At the outer surface of the oxide layer there is continual competition between the oxidizing action of the oxygen which tends to form more higher oxide and the reducing action of the ferro-ions which try to reduce the quantity of higher oxide. If the surrounding oxygen is pumped away before all the iron has been oxidized, the reducing action continues and after a certain time all the Fe_3O_4 and Fe_2O_3 will have been converted into Fe_xO.

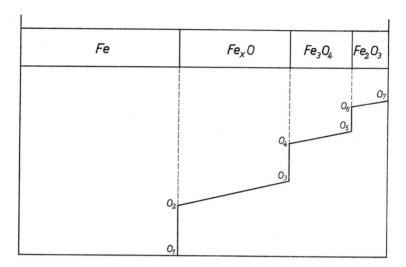

Fig. 12. Schematic representation of the oxidation of iron at e.g. 1000 °C under non-equilibrium conditions. If it is assumed that equilibrium is established at the interfaces, then the contents O_1, O_2 . . . correspond to those indicated by the same symbols in Fig. 10.

In the layers of Fe_xO, Fe_3O_4 and Fe_2O_3 (Fig. 12) the oxygen concentration increases from left to right, while the iron concentration decreases. The concentration curve for oxygen is schematically represented by the full line $O_1O_2O_3 \ldots O_7$ in Fig. 12. The concentrations at the interfaces, O_1, O_2, O_3, etc. can be approximately equated to the concentrations which can be read off from the phase diagram (Fig. 9). As already stated, we shall go into the kinetic aspects of the oxidation of metals more thoroughly in the second volume of this work [1] which has already been mentioned in the "General Introduction".

4.6. Reaction of iron and its oxides with mixtures of CO and CO₂ or H₂ and H₂O

The dissociation pressures of the iron oxides, which were discussed in the previous section, are too small to be measured directly. They can be determined by means of the equilibria

[1] J. D. FAST, *Interaction of Metals and Gases*, II. *Kinetics and Mechanisms*, to be published.

$$H_2 + \tfrac{1}{2} O_2 \rightleftharpoons H_2O, \qquad\qquad (4.6.1)$$

$$CO + \tfrac{1}{2} O_2 \rightleftharpoons CO_2, \qquad\qquad (4.6.2)$$

which are very accurately known from calorimetric and spectroscopic measurements. For (4.6.1) we have

$$K_p = \frac{p_{H_2O}}{p_{H_2} \cdot p_{O_2}^{\frac{1}{2}}},$$

and thus

$$p_{O_2} = \left(\frac{p_{H_2O}}{p_{H_2}}\right)^2 \cdot \frac{1}{K_p^2}. \qquad\qquad (4.6.3)$$

Rough values of K_p can be found with the help of Table 1 and equation (4.1.3). For 1000 °K one can calculate $K_p \simeq 10^{10}$. Substituting in (4.6.3):

$$p_{O_2}(1000\ °K) \simeq 10^{-20} \left(\frac{p_{H_2O}}{p_{H_2}}\right)^2.$$

With equal partial pressures of hydrogen and water vapour one thus automatically obtains a partial oxygen pressure of 10^{-20} atm at 1000 °K. The use of gas mixtures $(H_2 + H_2O)$ thus presents a fine method of adjusting very low, but accurately known oxygen pressures. The same is true for mixtures $(CO + CO_2)$. The reaction constant for the "water-gas-equilibrium"

$$CO_2 + H_2 \rightleftharpoons CO + H_2O \qquad\qquad (4.6.4)$$

is approximately equal to 1 at 1000 °K. At this temperature thus

$$\frac{p_{CO} \cdot p_{H_2O}}{p_{CO_2} \cdot p_{H_2}} \simeq 1$$

or

$$\frac{p_{CO}}{p_{CO_2}} \simeq \frac{p_{H_2}}{p_{H_2O}}.$$

According to this relationship a mixture $(H_2 + H_2O)$ is in equilibrium at 1000 °K with a $(CO + CO_2)$ mixture having the same pressure ratio. The two gas mixtures, which are in equilibrium with each other, naturally have the same equilibrium oxygen pressure.

Mixtures $(H_2 + H_2O)$, and thus also $(CO + CO_2)$, are particularly suitable for the investigation of the iron-oxygen system because the most interesting equilibria here do not lie, as was shown to be the case for $(Cu_2O + Cu)$ (see Section 4.2), entirely on the H_2O side, nor, as for

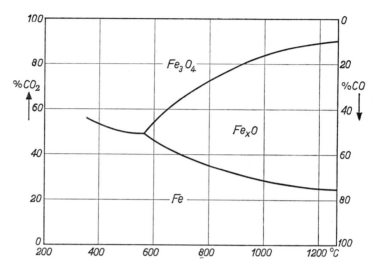

Fig. 13. Oxidation and reduction equilibria of iron and iron oxides with mixtures of CO and CO₂.

(Cr₂O₃ + Cr), entirely on the H₂ side. Figure 13 shows the positions of the equilibria with mixtures of CO and CO₂. The equilibrium

$$3\ Fe_2O_3 + CO \rightleftharpoons 2\ Fe_3O_4 + CO_2$$

does not appear in the figure since it lies so far on the CO₂ side that the curve representing it almost coincides with the upper horizontal axis.

Of the reactions which take place in the blast furnace, it is the reaction

$$Fe_{0.95}O + CO \rightleftharpoons 0.95\ Fe + CO_2 \tag{4.6.5}$$

which is of primary importance. The composition of the mixture (CO + CO₂) which is in equilibrium with (Fe₀.₉₅O + Fe) can be calculated to a very rough approximation with the help of Table 1 and equation (4.1.3). One finds

$$\log \frac{p_{CO_2}}{p_{CO}} \simeq \frac{800}{T} - 0.66.$$

For $T = 1000$ °K, this equation produces the value $p_{CO_2}/p_{CO} \simeq 1.38$, corresponding to 42% CO, 58% CO₂. If the temperature dependence of ΔH^0 and ΔS^0 is taken into account, as it should be, one arrives at a considerably higher percentage of CO. Precise measurements by Darken and Gurry show that in the temperature range 1000° -1600 °K the following relationship is true to a very good approximation:

$$\log \frac{p_{CO_2}}{p_{CO}} = \frac{850}{T} - 1.068. \tag{4.6.6}$$

At 1000 °K this formula gives $p_{CO_2}/p_{CO} = 0.61$, corresponding to 62% CO, 38% CO_2. At 1600 °K one finds $p_{CO_2}/p_{CO} = 0.29$, corresponding to 77% CO 23% CO_2. In agreement with these figures, it is found that the gas leaving the top of the blast furnace contains between two and three times as much CO as CO_2 (by volume). As was already discussed in Section 2.6, this proportion cannot be reduced by building taller blast furnaces. If the reaction

$$CO_2 + C \rightleftharpoons 2\,CO \tag{4.6.7}$$

is also taken into consideration, it may even seem surprising that the percentage of CO in the escaping gas is not higher still. In Fig. 14, curve d shows the position of equilibrium (4.6.7) at a pressure of 1 atm. Points on this curve give the composition of mixtures (CO + CO_2) at 1 atm which are in equilibrium with solid carbon. Points to the left of curve d relate to gas mixtures containing more CO than corresponds to equilibrium (4.6.7). In order to reach the equilibrium state at a particular temperature, CO must be converted into C and CO_2 until the point is reached on curve d which corresponds to that temperature. In general this decomposition of CO only occurs at an inappreciably low rate at low temperatures. The result of this

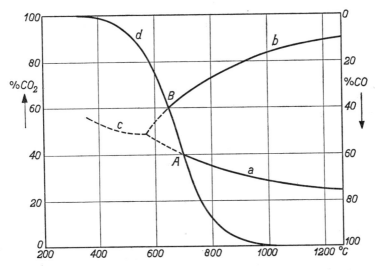

Fig. 14. The same curves as in Fig. 13, to which has been added the curve for the Boudouard-equilibrium, $CO_2 + C \rightleftharpoons 2\,CO$, for a pressure of 1 atm (curve d).

is that the equilibria, represented by points on the dotted sections of curves a, b and c, can be measured, although the corresponding mixtures ($CO + CO_2$) are not stable. They are, in fact, metastable equilibria.

Points in the space to the right of curve d represent gas mixtures with more CO_2 than is allowed by equilibrium (4.6.7). If sufficient carbon is present, CO_2 will react with it until the point is reached on curve d which corresponds to the prevailing temperature. If no carbon is present, the full lines of curves a and b represent stable equilibria. If the system contains solid carbon (as in a blast furnace), the equilibria represented by a and b will only be stable at points A and B. At point A, Fe, "FeO", C, CO and CO_2 are in equilibrium with one another.

At those temperatures at which equilibrium (4.6.7) is rapidly attained, the percentage of CO in the blast furnace gas is much higher than corresponds to (4.6.5). As a result, one might expect much higher percentages CO in the escaping gas than are actually found. If, on the other hand, equilibrium (4.6.7) were to be established rapidly at the relatively low temperatures encountered in the upper part of the blast furnace, then the escaping gas would be almost pure CO_2 (if the presence of nitrogen is neglected).

That CO must break down at low temperatures into C and CO_2 in order to establish the equilibrium state, can be deduced for a temperature of 25 °C from the ΔG^0 figures in Table 1.

4.7. The tungsten-oxygen and molybdenum-oxygen systems

As we have already seen, the phase equilibria in the iron-oxygen system correspond to easily-determined ratios H_2/H_2O and CO/CO_2 (see Fig. 13). They can therefore be studied conveniently with the aid of mixtures ($H_2 + H_2O$) or ($CO + CO_2$). The same is true for the phase equilibria which occur in the molybdenum-oxygen and tungsten-oxygen systems. There are, however, other metal-oxygen systems which are impossible or extremely difficult to study be means of these gas mixtures. As examples of these we discussed copper-oxygen and chromium-oxygen. Other examples are the binary systems of which oxygen is one component and vanadium, niobium, tantalum, titanium, zirconium or hafnium is the other. These systems also are unsuitable for investigation with mixtures ($H_2 + H_2O$) or ($CO + CO_2$) in the first place because the equilibrium ratios H_2/H_2O and CO/CO_2 are too large, and secondly because the six metals also have a great affinity for carbon and hydrogen (see Chapter 8).

Various investigators have applied the technique under consideration to

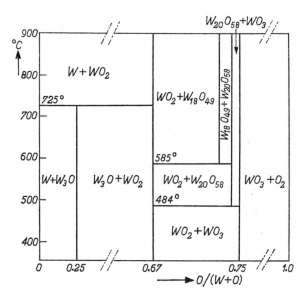

Fig. 15. The tungsten-oxygen phase diagram at one atmosphere total pressure according to St. Pierre and Speiser, Trans. AIME 224, 259 (1962). The numbers on the horizontal axis give the oxygen content (atom fraction).

the system W-O. These investigations form a welcome addition to the crystallographic investigations, carried out mainly by Hägg and Magnéli [1]). There are five oxides of tungsten: W_3O, WO_2, $W_{18}O_{49}$, $W_{20}O_{58}$ and WO_3. Fig. 15 shows the tungsten-oxygen phase diagram after St. Pierre and Speiser [2]), who made equilibrium measurements at various temperatures for the following reactions:

$$\tfrac{1}{2} WO_2 + H_2 \rightleftharpoons \tfrac{1}{2} W \quad + H_2O, \qquad (4.7.1)$$

$$\tfrac{1}{2} WO_2 + CO \rightleftharpoons \tfrac{1}{2} W \quad + CO_2, \qquad (4.7.2)$$

$$\tfrac{1}{13} W_{18}O_{49} + CO \rightleftharpoons \tfrac{18}{13} WO_2 \quad + CO_2, \qquad (4.7.3)$$

$$\tfrac{9}{32} W_{20}O_{58} + CO \rightleftharpoons \tfrac{10}{32} W_{18}O_{49} + CO_2, \qquad (4.7.4)$$

$$10\ WO_3 \quad + CO \rightleftharpoons \tfrac{1}{2} W_{20}O_{58} + CO_2. \qquad (4.7.5)$$

WO_3 exists in two or more enantiotropic crystalline forms, but the transition points are not shown on the phase diagram, nor are the homogeneity ranges of the oxides which have been only partially determined. The existence of the

[1]) G. Hägg and A. Magnéli, Revs. pure and appl. Chem. (Australia), 4, 235 (1954).
[2]) G. R. St. Pierre and R. Speiser, Trans. AIME 224, 259 (1962).

lowest oxide W_3O was first suggested by Hägg and Schönberg [1]). They presume that this is the composition of the phase "βW" which, for many years, had been regarded as an allotropic modification of tungsten. The opinion of Hägg and Schönberg is strongly supported by the X-ray work of Charlton and Davis [2]). There are indications that the phase under discussion has a homogeneity range extending to much smaller oxygen contents than that corresponding to the formula W_3O [3]).

Although the molybdenum-oxygen system has been less thoroughly investigated than the tungsten-oxygen system, it appears to be an established fact that the number of molybdenum oxides is even greater than the number of tungsten oxides [4–7]).

The solubility of oxygen in molybdenum and tungsten, like that in iron, is very small. According to Few and Manning [8]) the solubility limit of oxygen in molybdenum is 0.0045 weight percent at 1100 °C and 0.0065% at 1700 °C.

4.8. Aluminium and oxygen

In this section we shall consider the case of a metal which is liberated by electrolysis from the gas with which it is combined. We take as an example the commercially important manufacture of aluminium and ask what is the minimum voltage required to split a saturated solution of Al_2O_3 in molten cryolite, Na_3AlF_6, at 1000 °C into aluminium and oxygen at 1 atm.

The saturated solution is in equilibrium with solid Al_2O_3. The dissolved Al_2O_3 thus has the same oxygen dissociation pressure and also the same value of ΔG^0 as solid Al_2O_3. From Table 1 we calculate an approximate value of ΔG^0 at 1273 °K (1000 °C) by means of equation (4.1.1). For the reaction

$$\tfrac{4}{3}Al + O_2 \rightleftharpoons \tfrac{2}{3}Al_2O_3$$

the table gives

$$\Delta H^0 = -266 \text{ kcal},$$

$$\Delta S^0 = -49.5 \text{ cal/deg}.$$

[1]) G. Hägg and N. Schönberg, Acta Cryst. **7**, 351 (1954)
[2]) M. G. Charlton and G. L. Davis, Nature **175**, 131 (1955)
[3]) See also J. Bousquet and G. Pérachon, Compt. rend. **258**, 934 (1964)
[4]) A. Magnéli, Acta Chem. Scand. **2**, 501 and 861 (1948)
[5]) N. Schönberg, Acta Chem. Scand. **8**, 617 (1954)
[6]) L. Kihlborg and A. Magnéli, Acta Chem. Scand. **9**, 471 (1955)
[7]) L. Kihlborg, Arkiv för Kemi **21**, 471 (1963)
[8]) W. E. Few and G. K. Manning, Trans. AIME **194**, 271 (1952)

Since Al is liquid at 1000 °C, these values must be reduced respectively by the fusion enthalpy and fusion entropy of $\frac{4}{3}$ mole Al. These amount to 3 kcal and 3.5 cal/deg respectively [1]. Making use of these values, equation (4.1.1) gives:

$$\Delta G^0{}_{1273} \cong -269 + 1273 \times 0.053 = -202 \text{ kcal.}$$

With the help of equation (2.5.10) this gives us

$$E^0 \cong 2.2 \text{ volt.}$$

This voltage, at least, is necessary if the electrolysis is carried out with a platinum anode. Technically, carbon anodes are employed which react with the oxygen produced to form CO. The total result of the electrolysis is then given by the equation:

$$\tfrac{2}{3} Al_2O_3 + 2 C \rightleftharpoons \tfrac{4}{3} Al + 2 CO,$$

for which $\Delta G^0{}_{1273}$ is about 95 kcal, corresponding to a minimum voltage of about 1.0 volt. In reality one is forced to apply several volts when manufacturing Al, mainly as a result of the electrical resistance of the bath.

---

[1] cf. O. KUBASCHEWSKI and E. LL. EVANS, *Metallurgical Thermochemistry*, Pergamon Press, London, third edition, 1958.

CHAPTER 5

REACTIONS BETWEEN ALLOYS AND GASES WHEREBY NEW PHASES ARE FORMED

5.1. Ideal solutions

Table 1, discussed in Chapter 3, can give information about the position of the equilibrium between pure condensed substances and gases at not too high pressure. If the reaction partners also include substances which are dissolved in either a solid or liquid medium, the equilibrium can only be calculated from Table 1 if the solutions are ideal (see following paragraphs).

The vapour pressure of a component over a stable solution is always smaller than that of the pure component. A solution is said to obey Raoult's law when the vapour pressure of each component i is equal to its mole fraction x_i multiplied by its vapour pressure p_i^0 in the pure state:

$$p_i = x_i p_i^0. \tag{5.1.1}$$

The mole fraction x_i of a component i in a homogeneous mixture is the ratio of the number of moles of i to the total number of moles in the phase in question. In this section we shall only consider binary solutions and thus have:

$$\left. \begin{aligned} x_1 &= \frac{n_1}{n_1 + n_2}, \\ x_2 &= \frac{n_2}{n_1 + n_2}, \\ x_1 + x_2 &= 1 \ . \end{aligned} \right\} \tag{5.1.2}$$

A solution is considered ideal if (a) it satisfies Raoult's law, (b) no volume changes occur when the pure components are mixed, (c) the heat of mixing (enthalpy of mixing) is equal to zero. Ideal solutions are extremely rare. They are most closely approximated by mixtures of isotopes. The deviations from ideal behaviour are very small if the molar volumes of the components 1 and 2 are virtually equal and if, furthermore, the interaction energy between a pair of atoms or molecules 1-2 is virtually equal to the mean of the interaction energies of pairs 1-1 and 2-2.

For a binary, ideal solution Fig. 16 gives the partial vapour pressures p_1 and p_2 and the total vapour pressure

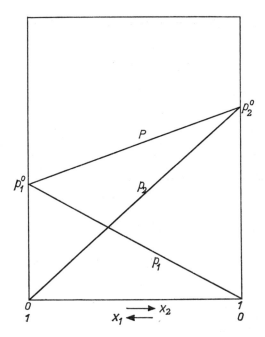

Fig. 16. If the two components of a binary system form an uninterrupted series of ideal solutions, then their vapour pressures p_1 and p_2 will be proportional to their mole fractions x_1 and x_2. The figure also gives the total vapour pressure P as a function of the concentration.

$$P = p_1 + p_2$$

as functions of the composition for a particular temperature; x_2 increases from left to right from 0 to 1, while x_1 increases from right to left from 0 to 1.

Once equilibrium has been established between liquid and vapour, then according to Section 2.2, the chemical potential of each component i in the solution will be equal to that of the same component in the vapour:

$$\mu_i^{sol} = \mu_i^{vap}. \qquad (5.1.3)$$

If we consider the vapour as a perfect gas, which is very often justified, relation (2.3.5) will be valid, so that (5.1.3) becomes:

$$\mu_i^{sol} = \mu_i^{vap} (1 \text{ atm}) + RT \ln p_i, \qquad (5.1.4)$$

in which p_i once more represents the partial pressure of i above the solution. The chemical potential of the vapour at 1 atm of substance i could not be indicated in (5.1.4) by the symbol μ_i^0, since this is reserved here for the chemical potential of the pure condensed substance i. Substitution of (5.1.1) in (5.1.4) gives:

$$\mu_i^{sol} = \mu_i^{vap} (1 \text{ atm}) + RT \ln p_i^0 + RT \ln x_i. \qquad (5.1.5)$$

According to (2.3.5),

$$\mu_i^{vap} (1 \text{ atm}) + RT \ln p_i^0$$

is none other than the μ of the saturated vapour above the pure substance i. This, in turn, is equal to μ_i^0 of the pure (condensed) substance i itself. Equation (5.1.5) thus becomes:

$$\mu_i^{sol} = \mu_i^0 + RT \ln x_i. \qquad (5.1.6)$$

The dependence of the chemical potential of a substance in ideal solution on its concentration (mole fraction) is seen from this equation to be wholly comparable with that of a perfect gas on its partial pressure:

$$\mu_i^{gas} = \mu_i^0 + RT \ln p_i. \qquad (5.1.7)$$

In (5.1.6) the symbol μ_i^0 refers to the pure condensed substance, in (5.1.7) to the pure gas at 1 atm.

5.2. The equilibrium constant of a reaction in which ideal solutions participate

In Section 2.4 we discussed the affinity and the equilibrium constant of reactions in which only gases and pure condensed substances took part. The starting point was the condition

$$\Delta G = \Sigma \nu_i \mu_i = 0 \qquad (2.2.7)$$

for chemical equilibrium. In this we inserted (2.3.5) for the participating gases and (2.3.6) for the pure condensed substances, so that the expression became

$$\Delta G^0 = -RT \ln \Pi \, p_j^{\nu_j}. \qquad (2.4.4)$$

If the reaction partners include not only gases and pure condensed substances but also substances in ideal solution, then substitution of (2.3.5), (2.3.6) and (5.1.6) in (2.2.7) immediately gives:

$$\Delta G^0 = -RT \ln \Pi \, p_j^{\nu_j} x_k^{\nu_k} . \qquad (5.2.1)$$

The equilibrium constant is thus given in this case by:

$$K_p = \Pi \, p_j{}^{\nu_j} x_k{}^{\nu_k}, \tag{5.2.2}$$

where the symbols p_j and ν_j refer to the gases and the symbols x_k and ν_k to the substances in ideal solution which take part in the reaction.

Let us take as an example the reaction

$$2 \, [\mathrm{Cr}] + \tfrac{3}{2} \, O_2 \rightleftharpoons (\mathrm{Cr_2O_3}),$$

where the square brackets indicate that Cr is one of the components of a homogeneous metallic phase and the round brackets that $\mathrm{Cr_2O_3}$ is one of the components of a homogeneous oxide phase. If both solutions may be regarded as ideal, then the following will be valid according to (5.2.2):

$$K_p = \frac{x'_{\mathrm{Cr_2O_3}}}{x_{\mathrm{Cr}}^2 \cdot p_{O_2}^{3/2}}.$$

where the symbol x refers to the metallic and the symbol x' to the oxide phase.

It will be seen from the derivation of formula (5.2.1) that the left-hand side, i.e.

$$\varDelta G^0 = \Sigma \nu_i \mu_i{}^0,$$

has the same significance as $\varDelta G^0$ in the previous chapter. It is the change in free enthalpy which would occur if the substances participating in the reaction were pure insofar as they are liquid or solid, and had a partial pressure of 1 atm (and furthermore were perfect) insofar as they are gaseous. The value of $\varDelta G^0$ can therefore once again be derived from Table 1.

It is obvious that the formulae from the previous chapter in which the reaction constant K_p occurs, are also applicable to the reactions discussed in this section in which ideal solutions participate. However, K_p must now be regarded as the quantity defined by formula (5.2.2). The important formula (4.1.3) follows, on this condition, directly from the formulae (5.2.1) and (4.1.1).

Before proceeding to a discussion of non-ideal solutions, we shall use the next two sections to demonstrate the application of the above by means of examples.

5.3. Selective oxidation

If oxygen or wet hydrogen reacts with a binary alloy, the components of which have very different affinities for oxygen, then under certain circumstances selective oxidation of the most reactive component may occur. If, for example, an alloy of iron and chromium is heated in wet hydrogen, then over a wide range of H_2/H_2O ratios there will only be oxidation of the chromium, which has a much greater affinity for oxygen than iron (see Table 1).

In this section we shall consider the oxidation of ideal solid solutions of two metals which have different affinities for oxygen and whose oxides are mutually insoluble and do not react to form compounds. Alloys of copper and nickel fulfill these conditions fairly well: to a rough approximation they can be considered as ideal solutions [1]) and the oxides Cu_2O and NiO do not dissolve in one another and form no compounds [2,3]). According to Table 1, nickel is the more reactive metal in the alloys in question. The change in free enthalpy which occurs during oxidation of the dissolved nickel according to the equation

$$[Ni] + \tfrac{1}{2} O_2 \rightleftharpoons NiO \qquad (5.3.1)$$

is seen from the above (neglecting the deviations from Raoult's law) to be given by

$$\Delta G = \Delta G^0 - RT \ln (p_{O_2}^{\frac{1}{2}} \cdot x_{Ni}), \qquad (5.3.2)$$

where ΔG^0 is the standard value of the free enthalpy of formation of NiO (Table 1). The change in free enthalpy which occurs during the oxidation of nickel in ideal solution by oxygen at 1 atm is therefore given by

$$\Delta G = \Delta G^0 - RT \ln x, \qquad (5.3.3)$$

where x is the mole fraction of nickel. The affinity for oxygen therefore decreases as the concentration of the dissolved element decreases. The affinities of [Cu] and [Ni] for oxygen are equal to one another when

$$\Delta G^0_{NiO} - RT \ln x = \Delta G^0_{Cu_2O} - RT \ln (1 - x)^2.$$

Making use of Table 1 and the approximation formula (3.2.9), one can calculate from this for a temperature of 1000 °K the value $x = 0.002$, which corresponds to 0.2% Ni. This result means that the nickel in a Cu-Ni alloy cannot be oxidized further than to a residual content of 0.2% without converting the copper into oxide at the same time. Elements with a greater affinity for oxygen than nickel, when dissolved in copper can in general be selectively oxidized up to a smaller residual content.

---

[1]) In reality solid Cu-Ni alloys exhibit positive heats of mixing and therefore definite deviations from Raoult's law. See: R. A. RAPP and F. MAAK, Acta Met. **10**, 63 (1962).

[2]) J. A. SARTELL, S. BENDEL, T. L. JOHNSTON and C. H. LI, Trans. Amer. Soc. Metals **50**, 1047 (1958).

[3]) N. G. SCHMAHL and F. MÜLLER, Z. anorg. allg. Chem. **332**, 217 (1964).

The residual content of Ni can also be calculated directly by means of the equilibrium

$$[Ni] + Cu_2O \rightleftharpoons NiO + 2 \, [Cu]. \qquad (5.3.4)$$

The equilibrium constant

$$K_p = \frac{(1-x)^2}{x}$$

is found approximately for 1000 °K with the help of equation (4.1.3). Again one finds $x = 0.002$ (0.2 %Ni).

For the production of copper with a high electrical conductivity, the above-mentioned selective oxidation can be used to advantage by converting impurities such as Fe, As, Sb and Ni, which increase the electrical resistance, into oxides with the aid of dissolved oxygen (see Section 4.3). This internal oxidation reduces the electrical resistance of the metal [1].

It is obvious that the fact that nickel becomes less reactive as its solution in copper becomes more dilute, must also appear in an increased oxygen dissociation-pressure of NiO which is in equilibrium with dissolved Ni. Consider the dissociation reactions

$$2 \, NiO \rightleftharpoons 2 \, Ni + O_2 \qquad (5.3.5)$$

$$2 \, NiO \rightleftharpoons 2 \, [Ni] + O_2. \qquad (5.3.6)$$

The reaction constants are given by

$$K(5) = p_{O_2}^0,$$

$$K(6) = p_{O_2} \cdot x_{Ni}^2.$$

According to the previous section $K(5)$ and $K(6)$ are equal, so that

$$p_{O_2}^0 = p_{O_2} \cdot x_{Ni}^2. \qquad (5.3.7)$$

Since x is always smaller than 1, p_{O_2} is always larger than $p_{O_2}^0$ and will exceed it more as x becomes smaller.

In the Sections 5.6 et seq. it will be seen that the results found here are more generally valid and can also be applied to non-ideal solutions: If a metal is alloyed with one or more metals which are less reactive, it also becomes less reactive (more noble), provided that a homogeneous solution is formed. This is shown in two ways: firstly, the dissolved metal is less easily oxidized than the pure metal, secondly, its oxide is more easily reduced if the metal formed is dissolved by less reactive metals.

---

[1] F. PAWLEK and collaborators, Z. Metallkunde **47**, 347 and 357 (1956).

In the following section we shall discuss a more complicated case of selective oxidation than in this section.

5.4. Oxidation of iron containing manganese

In alloys of iron and manganese, the latter is the more reactive component (Table 1). The oxidation of iron-manganese alloys differs from that of copper-nickel alloys in that the oxides formed, FeO and MnO (we neglect deviations from stoichiometric composition), form an uninterrupted series of solid solutions. These can be regarded as ideal solutions to a very good approximation [1]). At high temperatures (e.g. 1150 °C) it is probable that also Fe and Mn form homogeneous solid solutions in all or nearly all proportions. According to Butler et al. [2]) these solutions exhibit only small departures from ideality. We shall regard them in this section as ideal solutions because the divergences from ideal behaviour make no essential difference to the important conclusions to which a simple calculation will lead us. Without introducing great errors we can, further, presume that no oxide or oxygen dissolves in the metallic phase [Fe + Mn] and no metal or oxygen in the oxide phase (FeO + MnO).

A good impression of the selectivity of the oxidation is obtained by considering the equilibria between solid solutions of FeO and MnO on one hand and alloys of Fe and Mn on the other hand. The reaction which governs these equilibria can be written

$$(MnO) + [Fe] \rightleftharpoons [Mn] + (FeO), \qquad (5.4.1)$$

where the square brackets as usual indicate a component of a homogeneous metallic phase and the round brackets a component of a homogeneous oxide phase. Since the solutions are to be regarded as ideal, the equilibrium constant for this reaction is given, according to (5.2.2), by

$$K = \frac{x_{Mn} \cdot x'_{FeO}}{x_{Fe} \cdot x'_{MnO}}, \qquad (5.4.2)$$

where the symbol x refers to the metallic phase and the symbol x' to the oxide phase. If x_{Mn} is indicated simply by x, then $x_{Fe} = (1 - x)$. If, furthermore, we abbreviate x'_{MnO} to x', then $x'_{FeO} = (1 - x')$. We thus obtain

---

[1]) P. K. FOSTER and A. J. E. WELCH, Trans. Faraday Soc. **52**, 1636 (1956); H. SCHENCK, N. G. SCHMAHL and A. K. BISWAS, Arch. Eisenhüttenwes. **28**, 517 (1957)
[2]) J. F. BUTLER, C. L. MCCABE and H. W. PAXTON, Trans. AIME **221**, 479 (1961)

$$K = \frac{x\,(1-x')}{x'\,(1-x)}. \tag{5.4.2a}$$

According to (5.2.1), the value of K is given by

$$\Delta G^0 = -RT \ln K$$

or

$$\log K = -\frac{\Delta H^0}{4.575\,T} + \frac{\Delta S^0}{4.575},$$

where ΔG^0, ΔH^0 and ΔS^0 refer to the reaction between the pure (not dissolved) substances. We wish to find K at e.g. 1150 °C and calculate an approximate value of this quantity in the familiar manner by inserting in the last equation the values derived from Table 1 for ΔH^0 and ΔS^0. It can easily be verified that the following value is found in this way:

$$K\,(1150\ ^\circ\mathrm{C}) = \frac{x\,(1-x')}{x'\,(1-x)} \cong 4 \times 10^{-5}. \tag{5.4.3}$$

Several corresponding values of x and x', which have been found with the help of this equation, are given below. Naturally, only manganese-free oxide $(x'_{\mathrm{MnO}} = x' = 0)$ can form on pure iron $(x_{\mathrm{Mn}} = x = 0)$.

$x_{\mathrm{Mn}} = x$	% Mn in the metal	$x'_{\mathrm{MnO}} = x'$	% MnO in the oxide
0	0	0	0
10^{-5}	10^{-3}	0.20	20
$4 \cdot 10^{-5}$	$4 \cdot 10^{-3}$	0.50	50
10^{-4}	10^{-2}	0.71	71
10^{-3}	10^{-1}	0.96	96
10^{-2}	1	0.996	99.6

Since equation (5.4.3) is symmetrical, we find exactly the same pairs of corresponding values for $(1-x')$ and $(1-x)$:

$x'_{\mathrm{FeO}} = (1-x')$	% FeO in the oxide	$x_{\mathrm{Fe}} = (1-x)$	% Fe in the metal
0	0	0	0
10^{-5}	10^{-3}	0.20	20
$4 \cdot 10^{-5}$	$4 \cdot 10^{-3}$	0.50	50
10^{-4}	10^{-2}	0.71	71
10^{-3}	10^{-1}	0.96	96
10^{-2}	1	0.996	99.6

The equilibria calculated express the relationships in a portion of the ternary diagram iron-manganese-oxygen and can thus be graphically represented in part of a triangle, the corners of which correspond to Fe, Mn and $\frac{1}{2}$ O$_2$ (Fig. 17). A clearer diagram is obtained by representing the equilibria which interest us in a square, the corners of which refer to Fe, Mn, FeO and MnO (Fig. 18). Horizontally x or x' are plotted, vertically the ratio y of oxygen atoms to metal atoms. According to the premises, the single-phase regions are restricted to the horizontal sides Fe-Mn ($y = 0$) and FeO-MnO ($y = 1$). The extremities of the sloping lines in Fig. 18 represent the calculated compositions of the metallic and oxide phases which can co-exist. These lines are known as tie-lines. In the case under discussion they are given by equation (5.4.3). A mixture corresponding to an arbitrary point P in the square, separates into two phases, the compositions of which are given by the ends of the tie-line passing through P. The quantities of the two phases are inversely proportional to the distances from P to the two ends.

As follows from the above, the equilibria discussed can be divided, roughly speaking, into two groups. One is either dealing with a solid solution of FeO and MnO in equilibrium with almost pure iron, or with a solid solution of Fe and Mn in equilibrium with almost pure MnO. Thus, if an alloy of Fe and Mn is partially oxidized, principally MnO will be formed. The tie-lines representing this situation radiate from the MnO corner to points on the Fe-Mn line. Conversely, if a solid solution of FeO and MnO is partially

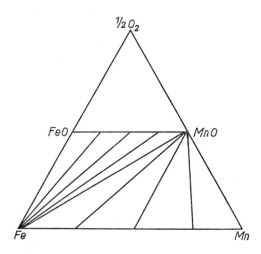

Fig. 17. In the Fe-Mn-FeO-MnO portion of the ternary diagram iron-manganese-oxygen the tie lines take the form shown in the figure.

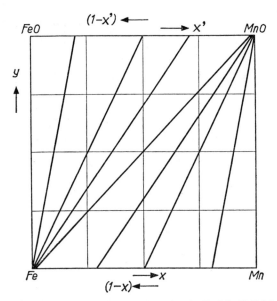

Fig. 18. Representation in a square of the tie lines in the Fe-Mn-FeO-MnO portion of the ternary diagram iron-manganese-oxygen.

reduced, almost pure iron will be formed. The tie-lines representing this, diverge from the Fe corner towards points on the FeO-MnO line.

In reality, as can be seen from the calculations, the two bundles of lines do not pass exactly through the corners Fe and MnO. Fig. 19 shows in a schematic manner the paths of the tie-lines for a system X-Y-XO-YO, in which the stabilities of the oxides differ less than is the case for FeO and MnO.

Since reaction (5.4.1) is endothermic, the reaction constant (5.4.2), according to Section 4.2, will increase with increasing temperature, i.e. the equilibrium will shift to the right. This is of great importance in steelmaking. With the help of Table 1 we calculate the value of K for a temperature of 1600 °C, at which the two phases are liquid. If it is assumed that not only the specific heats, but also the heats of fusion, cancel one another on the two sides of the equation, one finds for this temperature:

$$K = 5 \times 10^{-4}.$$

Values have been found experimentally which are larger by a factor of about 6 [1]). This is hardly surprising if one takes into account all that has

[1]) C. WAGNER, *Thermodynamics of Alloys*, Addison-Wesley Press, Cambridge, Mass., 1952.

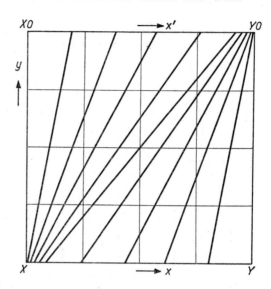

Fig. 19. In Figs 17 and 18 the two bundles of tie lines seem to pass through the points corresponding to Fe or MnO. If the affinities of iron and manganese for oxygen differed only a little from one another, the tie lines would be disposed as shown in this figure.

been neglected and approximated. Application of the more accurate method of calculation, described in Section 4.4, gives better agreement with the experimental results.

5.5. Chemical reactions in which non-ideal solutions participate

Equation (5.1.1) is not valid for non-ideal solutions: The ratio, p_i/p_i^0, of the vapour pressure p_i of a substance i above a solution to the vapour pressure p_i^0 of the pure substance i, does *not*, in general, correspond to the mole fraction x_i of i in the solution. The ratio p_i/p_i^0 is called the *activity* a_i of component i:

$$p_i = a_i p_i^0. \tag{5.5.1}$$

Equation (5.5.1) gives the definition of the quantity a and is therefore generally valid.

The manner in which equation (5.1.6) was derived for an ideal solution can be used to find a relationship of more general validity:

$$\mu_i^{\text{sol}} = \mu_i^0 + RT \ln a_i. \tag{5.5.2}$$

It is also immediately obvious that equations (5.2.1) and (5.2.2) become the more general formulae

$$\Delta G^0 = -RT \ln \Pi \, p_j{}^{\nu_j} a_k{}^{\nu_k} \tag{5.5.3}$$

and

$$K_p = \Pi \, p_j{}^{\nu_j} a_k{}^{\nu_k}, \tag{5.5.4}$$

where the index j again refers to the gases and the index k to the substances in solution which participate in the reaction. As an example we consider the reaction

$$(FeO) + CO \rightleftharpoons [Fe] + CO_2, \tag{5.5.5}$$

where the FeO is supposed to be dissolved in a slag and the Fe in a homogeneous alloy. In this case the equation

$$K_p = \frac{a_{Fe} \cdot p_{CO_2}}{a_{FeO} \cdot p_{CO}} \tag{5.5.6}$$

is always correct, however large are the deviations from ideality. ΔG^0 and K_p once more have the *same* values as for the reaction corresponding to (5.5.5) in which pure FeO and pure Fe take part. The equilibrium ratio p_{CO_2}/p_{CO}, however, in general has a different value, since the quotient a_{Fe}/a_{FeO} is only equal to 1 in exceptional cases.

Unfortunately, equations (5.5.3) and (5.5.4) can only be usefully employed when the experimental relation between a and x is known. The ratio a/x is called the activity coefficient f:

$$a_i = f_i x_i. \tag{5.5.7}$$

Equation (5.5.1) shows that for a pure substance a and x have the same value: $a = x = 1$. The activity defined by (5.5.1) is therefore called the activity with respect to the pure substance as reference state. The reference state here coincides with the standard state of a liquid or solid substance as defined in Section 2.4 and which is sometimes called the standard reference state. The choice of the pure substance as the reference state is particularly convenient for systems which do not depart far from ideal behaviour and where, thus, the activity coefficients are correction factors which differ comparatively little from unity. This is the case, for example, in many liquid and solid alloys which form an uninterrupted series of homogeneous solutions.

Typical curves for the vapour pressure of a substance 2 above binary solutions of substances 1 and 2 are to be found in Fig. 20. By adjusting the units on the vertical axis, these curves are also the activity curves for sub-

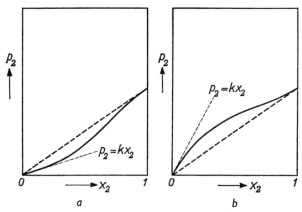

Fig. 20. If the two components of a binary system form an uninterrupted series of non-ideal solutions, then the vapour pressure of the components varies in the manner shown schematically for one of them. Negative deviations from Raoult's Law (*a*) occur, in general, when mixing is an exothermal process, positive deviations (*b*) when mixing occurs endothermally.

stance 2, since the activities are found, according to (5.5.1), by dividing the vapour pressures by a constant. In general one finds negative deviations from Raoult's law (Fig. 20a) when heat is liberated on mixing the components. For example, this is the case in the liquid systems Ag-Cd, Ag-Mg, Au-Zn, Cu-Zn, Hg-K, Hg-Na and Hg-Tl. In contrast, positive deviations are observed if mixing is an endothermic process (absorption of heat during isothermal mixing). This is the case, for example, for the liquid systems Pb-Cd, Cd-Sn, Cd-Zn, Bi-Hg, Pb-Hg, Sn-Hg and Sn-Zn.

5.6. Determination of activities by measurement of gas pressures or electromotive forces

In many cases it is possible, by measurement of gas pressures, to determine the relationship between the concentration and the activity of a dissolved metal. To illustrate this we shall discuss the determination of the activity of silver in silver-gold alloys by measurement of the dissociation pressure of silver chromate Ag_2CrO_4. At high temperatures, in an enclosed space, this compound decomposes partially into silver, oxygen and silver chromite $Ag_2Cr_2O_4$ [1]:

$$Ag_2CrO_4 \rightleftharpoons \tfrac{1}{2} Ag_2Cr_2O_4 + Ag + O_2. \qquad (5.6.1)$$

[1] R. SCHENCK, A. BATHE, H. KEUTH and S. SÜSS, Z. anorg. allg. Chem. **249**, 88 (1942); N. G. SCHMAHL, Z. anorg. allg. Chem. **266**, 1 (1951).

The oxygen pressures in this equilibrium above 400 °C are so large that they can be measured directly without difficulty. If finely divided gold is added to the silver chromate then, when decomposition takes place at e.g. 550 °C, homogeneous solid solutions of silver and gold are formed and the oxygen dissociation pressures rise. Reaction (5.6.1) must now be written

$$Ag_2CrO_4 \rightleftharpoons \tfrac{1}{2} Ag_2Cr_2O_4 + [Ag] + O_2. \tag{5.6.2}$$

The equilibrium constants of (5.6.1) and (5.6.2) are given by

$$K(1) = p_{O_2},$$

$$K(2) = p'_{O_2} \cdot a_{Ag}.$$

Since the constants are equal, we obtain:

$$a_{Ag} = \frac{p_{O_2}}{p'_{O_2}}. \tag{5.6.3}$$

Since a_{Ag} for reaction (5.6.2) is always smaller than 1, p'_{O_2} is always greater than p_{O_2}. By measuring these pressures, it is possible to determine a_{Ag} for various compositions of the Ag-Au alloy, i.e. for various values of x_{Ag}. Fig. 21 shows the relationship between a_{Ag} and x_{Ag}, found in this way for a temperature of 550 °C.

Activities of silver in silver-gold alloys have also been determined in a completely different way, viz. by measurement of electromotive forces in suitably chosen galvanic cells. The electrolyte in one of these cells is a molten mixture of KCl and AgCl, which can transport silver ions. Rods of pure silver and of a silver-gold alloy are used as anode and cathode. At the silver, the following reaction takes place

$$Ag \rightarrow Ag^+ + e^-. \tag{5.6.4}$$

The electrons liberated flow through the external circuit to the cathode. They react there with silver ions, forming silver atoms which are incorporated into the silver-gold alloy:

$$Ag^+ + e^- \rightarrow [Ag]. \tag{5.6.5}$$

Fig. 22 gives a schematic diagram of the cell described above. The result of the "half cell reactions" (5.6.4) and (5.6.5) is the "overall cell reaction"

$$Ag \rightarrow [Ag], \tag{5.6.6}$$

which, according to equations (2.2.5) and (5.5.2), is accompanied by a change

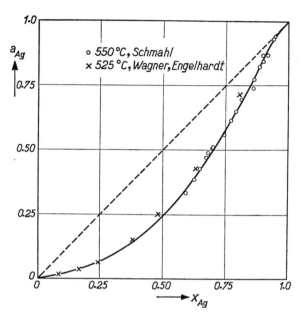

Fig. 21. Activities of silver at 525° and 550 °C in alloys of silver and gold calculated from the dissociation pressures of the reactions (5.6.1) and (5.6.2) by SCHMAHL, Z. anorg. allg. Chem. **266**, 1 (1951) and from electromotive forces by WAGNER and ENGELHARDT, Z. phys. Chem. A **159**, 241 (1932).

in the free enthalpy of $RT \ln a$. From Section 2.5, this change in free enthalpy is also given by $-zFE$, in which E is the electromotive force of the cell. Since $z = 1$, the relationship

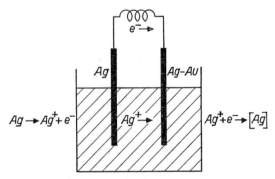

Fig. 22. The activity of silver in a silver-gold alloy can be found by measuring the electromotive force of a galvanic cell, the electrodes of which are made of pure silver and the alloy concerned.

$$-FE = \Delta G = RT \ln a$$

is valid in this particular case, so that measurement of the electromotive force gives directly the activity of the silver in the silver-gold alloy. By using alloys of various compositions as cathode, Wagner and Engelhardt [1] determined activities of silver in silver-gold alloys at 525 °C which agree excellently with the values obtained by the measurement of gas pressures (see Fig. 21).

5.7. Alteration of the reducibility of metal compounds by means of added substances

Most metals are won by reduction of oxide ore or by reduction of oxides obtained by roasting. As a measure of the reducibility of these substances, the oxygen dissociation pressure is often chosen. The greater this pressure, the better the reducibility. According to old experience of metallurgists, the reducibility of oxides which were difficult to reduce could, in many cases, be improved by adding more easily reducible oxides. Conversely, a decrease in the reducibility of oxides often occurs when less easily reducible oxides are added. One of the conditions for the occurrence of the former phenomenon is the mutual solubility of the two metals to be formed. The latter case requires the oxides to be mutually soluble or to form a compound. This will be further explained below, using the oxide of a bivalent metal as an example. The discussion can be extended without difficulty to metals of any valency.

For the dissociation of the pure oxide we have:

$$2\text{ MeO} \rightleftharpoons 2\text{ Me} + O_2; \quad K = p_{O_2}^0. \tag{5.7.1}$$

Now suppose that we add to this oxide a more easily reducible oxide and that the two oxides, for either kinetic or thermodynamic reasons, do not dissolve in one another. During the reduction of the mixture of oxides, the metal corresponding to the more easily reducible oxide will be formed first. We shall assume that metal Me dissolves in it. In that case, the symbol Me in (5.7.1) will appear in square brackets and the reaction constant will be given by

$$K = p_{O_2} \cdot a_{Me}^2. \tag{5.7.2}$$

Combination of (5.7.1) and (5.7.2) gives:

---

[1] C. WAGNER and G. ENGELHARDT, Z. physik. Chem. A. **159**, 241 (1932).

$$p^0_{O_2} = p_{O_2} \cdot a^2_{Me}. \qquad (5.7.3)$$

From this follows that $p_{O_2} > p^0_{O_2}$, i.e. improved reducibility of MeO by the addition of the more easily reducible oxide.

Matters are reversed if, to the oxide to be reduced one adds another, less-easily reducible oxide, in which the former dissolves. If the required metal, Me, is formed in the pure state by the reduction, then reaction (5.7.1) must be written

$$2\,(MeO) \rightleftharpoons 2\,Me + O_2. \qquad (5.7.4)$$

The reaction constant is given by

$$K = \frac{p'_{O_2}}{a^2_{MeO}}. \qquad (5.7.5)$$

Combination of (5.7.1) and (5.7.5) gives:

$$p^0_{O_2} \cdot a^2_{MeO} = p'_{O_2}.$$

From this follows that $p'_{O_2} < p^0_{O_2}$ and thus a decreased reducibility of MeO as a result of the addition of the less-easily reducible oxide.

The first case, the increased reducibility, is characterized by the conditions $a_{MeO} = 1$, $a_{Me} < 1$, the second, the decreased reducibility, by the conditions $a_{Me} = 1$, $a_{MeO} < 1$. The example discussed in the previous section is a variant of the first case. What we have dealt with in this section is not only applicable to oxides, but also, for example, to chlorides, sulphides and carbides. The next section contains further examples. They have been borrowed from the work of Schenck and Schmahl [1]).

5.8. Examples of shifting of metal-gas equilibria by added substances

The reaction which describes the roasting of copper

$$Cu_2S + 2\,Cu_2O \rightleftharpoons 6\,Cu + SO_2 \qquad (5.8.1)$$

can be strongly influenced by the addition of gold. Without this addition the SO_2 dissociation pressure is 118 mm Hg at 610 °C. If one adds a quantity of gold such that the SO_2 is in equilibrium with a copper-gold alloy containing 82.3 atom% Cu, then the SO_2 pressure is 500 mm. In a manner analogous to that above, one obtains the relationship

---

[1]) R. SCHENCK, Angew. Chem. **49**, 649 (1936); N. G. SCHMAHL, Z. anorg. allg. Chem. **266**, 1 (1951) and Angew. Chem. **65**, 447 (1953)

$$p_{SO_2} = p'_{SO_2} \cdot a_{Cu}^6. \tag{5.8.2}$$

From this one can directly calculate the activity of copper in a Cu-Au alloy with 82.3 atom % Cu: $a_{Cu} = 0.786$. There is thus a clear negative deviation from Raoult's law (see Section 5.5). As the gold content increases the SO_2 pressure increases much more rapidly than the O_2 pressure in the case of the dissociation of silver chromate (Section 5.6). The reason for this is the fact that a_{Cu} appears in (5.8.2) with the exponent 6, while a_{Ag} in (5.6.3) has the exponent 1.

The reverse, i.e. the shift of a reduction equilibrium to the left, occurs in the case of the reduction of silver sulphide by hydrogen

$$Ag_2S + H_2 \rightleftharpoons 2\,Ag + H_2S, \tag{5.8.3}$$

if Cu_2S is added to the Ag_2S. These substances form an uninterrupted series of solid solutions. Without additions one has

$$K = \frac{p_{H_2S}}{p_{H_2}}, \tag{5.8.4}$$

and with addition

$$K = \frac{p'_{H_2S}}{p'_{H_2}} \cdot \frac{1}{a_{Ag_2S}}.$$

The percentage H_2S in the equilibrium gas will thus drop, due to the addition of Cu_2S. The magnitude of this drop has been used to calculate the activities of Cu_2S and Ag_2S in their solutions.

A decrease in the reducibility of an oxide by the addition of a less-easily reducible one, not only occurs if the two oxides are mutually soluble, but also if they form a stable compound. To quote an example: when SiO_2 is added to FeO, the compound Fe_2SiO_4 is formed at high temperatures. It is much more difficult to reduce than FeO. The position of the new equilibrium

$$\tfrac{1}{2}\,Fe_2SiO_4 + CO \rightleftharpoons Fe + \tfrac{1}{2}\,SiO_2 + CO_2$$

can easily be calculated to an approximation by the reader with the help of Table 1.

ATTACK OF COPPER-GOLD ALLOYS BY OXYGEN

A particularly interesting case of selective oxidation is the interaction of oxygen and copper-gold alloys. At low gold contents a film of Cu_2O is formed on the alloys, while at high gold contents a film of CuO is formed. In order to understand this we shall consider the dissociation equilibria

$$2\,Cu_2O \rightleftharpoons 4\,Cu + O_2, \tag{5.8.5}$$

$$2\,CuO \rightleftharpoons 2\,Cu + O_2. \tag{5.8.6}$$

At 1000 °C, Cu_2O has a dissociation pressure of about 0.004 mm, CuO about 0.6 mm. According to the above, both dissociation pressures increase if the copper is diluted with gold. The new pressures are given by:

$$p_{O_2}(Cu_2O) = \frac{p^0_{O_2}(Cu_2O)}{a^4_{Cu}}, \tag{5.8.7}$$

$$p_{O_2}(CuO) = \frac{p^0_{O_2}(CuO)}{a^2_{Cu}}. \tag{5.8.8}$$

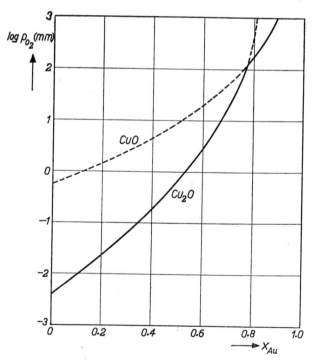

Fig. 23. During the oxidation of copper-gold alloys a layer of Cu_2O forms on the metal when the gold content is smaller than 77 atom %, while CuO is formed when the gold content exceeds this figure.

It can be seen that as a_{Cu} diminishes, the oxygen pressure above Cu_2O increases more rapidly than above CuO. At a copper content of about 23 atom %, the dissociation pressure of Cu_2O in equilibrium with the Cu-Au alloy has caught up with that of CuO (see Fig. 23). During the oxidation of copper-gold alloys under equilibrium conditions a layer of Cu_2O will thus form on the surface if the alloy contains more than 23 atom % Cu, while a layer of CuO will form if this content lies below 23 atom %.

CHAPTER 6

REACTION BETWEEN CARBON AND OXYGEN IN STEEL

6.1. Activity of carbon in iron

One of the most important metal-gas reactions is that between carbon and oxygen dissolved in liquid iron, producing the gases carbon monoxide and carbon dioxide. We shall discuss this reaction thoroughly, not only because of its technical importance, but also because it offers an opportunity to elucidate various aspects of the thermodynamics of metal-gas reactions. As an introduction to the discussion of the reaction in question we shall consider in this section the activity of carbon in iron and in the following section that of oxygen in iron.

Solutions of carbon in solid iron [1,2] and in liquid iron [3,4] have been studied by bringing the metal at various temperatures into equilibrium with mixtures of CO and CO_2 or CH_4 and H_2. If we take the first pair of gases as an example, then the following is valid for the equilibrium with *graphite* at a particular temperature

$$C + CO_2 \rightleftharpoons 2\,CO; \quad K = \frac{p_{CO}^2}{p_{CO_2}}. \tag{6.1.1}$$

For the equilibrium with a solution of carbon in iron we have

$$[C] + CO_2 \rightleftharpoons 2\,CO; \quad K = \frac{p_{CO}^2}{p_{CO_2} \cdot a_C}. \tag{6.1.2}$$

If graphite is chosen as the reference state of carbon, then the constant K in (6.1.2) has the same value as K in (6.1.1). The activity of carbon in iron with respect to graphite is thus given by

$$a_C = \frac{(p_{CO}^2/p_{CO_2}) \text{ in equilibrium with } [C] \text{ in Fe}}{(p_{CO}^2/p_{CO_2}) \text{ in equilibrium with graphite}}. \tag{6.1.3}$$

[1] H. Dünwald and C. Wagner, Z. anorg. allg. Chem. 199, 321 (1931)
[2] R. P. Smith, J. Amer. Chem. Soc. 68, 1163 (1946)
[3] F. D. Richardson and W. E. Dennis, Trans. Far. Soc. 49, 171 (1953)
[4] A. Rist and J. Chipman, *The Physical Chemistry of Steelmaking*, Edited by J. F. Elliott, Massachusetts, 1958 (p. 3)

The denominator of this fraction (i.e. K) has a value which has been experimentally determined for temperatures below 1000 °C and which, furthermore, (also for higher temperatures) can be accurately calculated from calorimetric and spectroscopic data [1]). At temperatures in the neighbourhood of 1800 °K, K is given by

$$\log K = -\frac{8435}{T} + 8.835 .$$ (6.1.4)

It is important to keep in mind that this equation not only applies to the Boudouard reaction, as the reaction which appears in equation (6.1.1) is often called, but also to the reaction given by equation (6.1.2).

By means of (6.1.4) the denominator of (6.1.3) can be calculated for various temperatures. The activity corresponding to a known concentration of carbon in iron can then be determined for each of these temperatures with the help of (6.1.3), by investigating which mixture of CO and CO_2 is in equilibrium with the Fe-C solution in question. As an illustration, Fig. 24 gives some experimental results of Richardson and Dennis (l.c.) from which the activity of carbon in liquid iron at 1560 °C can be derived. For carbon contents smaller than 0.3 weight% (1.5 atom%), the activity is seen from the figure to be roughly proportional to the concentration. This is often

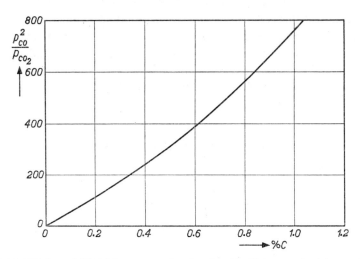

Fig. 24 Equilibria at 1560 °C between carbon-bearing liquid iron and mixtures of CO and CO_2 according to RICHARDSON and DENNIS, Trans. Faraday Soc. **49**, 171 (1953). The carbon content on the horizontal axis is given in wt%.

---

[1]) See e.g. J. D. FAST, Philips Res. Rep. **2**, 205 (1947)

expressed by saying that in this region the law of Henry is approximately valid. Strictly speaking, this law is only valid for infinitely dilute solutions (see the dotted tangents in Fig. 20). At higher contents than 0.3 weight % C, Fig. 24 shows definite deviations from Henry's law.

According to equation (5.5.2), the chemical potential of carbon in the solutions is given by

$$\Delta \mu_C = RT \ln a_C, \tag{6.1.5}$$

where $\Delta \mu_C = \mu_C - \mu^0(\text{graphite})$. In other words, equation (6.1.5) gives the chemical potential (partial molar free enthalpy) of carbon with respect to graphite as zero level. According to Section 2.2 it is given by

$$\Delta \mu_C = \Delta \bar{h}_C - T \Delta \bar{s}_C, \tag{6.1.6}$$

where $\Delta \bar{h}_C$ and $\Delta \bar{s}_C$ represent the partial molar enthalpy and the partial molar entropy of carbon in iron with respect to graphite. In other words, these are the changes in enthalpy and entropy accompanying the dissolution of 1 mole graphite in a quantity of an Fe-C alloy so large that the concentration remains virtually unchanged.

In general, $\Delta \bar{h}$ and $\Delta \bar{s}$ are dependent on both the temperature and the concentration. Their values in the example under discussion can be determined by plotting $\log a_C$ against $1/T$ for a number of constant carbon concentrations. If this is done one finds for both face-centred cubic iron (austenite or γ iron) and liquid iron mutually parallel straight lines. The enthalpy of solution $\Delta \bar{h}_C$ can therefore be regarded as virtually independent of temperature and concentration (see Table 10). On the other hand, as would be expected, the entropy of solution $\Delta \bar{s}_C$ becomes larger as the carbon concentration becomes smaller.

TABLE 10

HEAT OF SOLUTION IN CAL/MOLE AND ENTROPY OF SOLUTION IN CAL/DEG·MOLE OF
GRAPHITE IN FACE-CENTRED CUBIC IRON AND IN LIQUID IRON [1].

weight % C	austenite		liquid iron	
	$\Delta \bar{h}_C$	$\Delta \bar{s}_C$	$\Delta \bar{h}_C$	$\Delta \bar{s}_C$
1	10,800	9.75	6,400	9.70
0.5	10,800	11.48	6,400	11.50
0.3	10,800	12.58	6,400	12.62
0.1	10,800	14.78	6,400	14.91
0.01	10,800	19.35	6,400	19.48
0.001	10,800	23.93	6,400	24.06
0.0001	10,800	28.50	6,400	28.63

[1] F. D. RICHARDSON, J. Iron Steel Inst. **175**, 33 (1953).

The values of $\Delta \bar{s}_C$ for the three smallest carbon concentrations in Table 10 were obtained by extrapolation according to Henry's law, i.e. by assuming that the activity of carbon at small concentrations is proportional to the concentration. In this way, even in the most unfavourable case, only very small errors can be introduced.

The partial molar entropy of solution of one of the components in a homogeneous mixture in the simplest case, viz. that of ideal and "regular" solutions [1]), is given by $-R \ln x$, where x is the mole fraction of the component in question (cf. Section 5.1). This involves the assumption that this component has the same structure as the solution, thus, for example, is liquid in the case of liquid solutions. Graphite has a different structure from its solutions in iron. If they behaved as regular solutions, then the partial molar entropy of solution of graphite would be given by $-R \ln x$ plus a transition entropy (the fusion entropy of graphite in the case of the liquid solutions).

For the four smallest carbon concentrations (Table 10) we can write

$$\Delta \bar{s}_C = -R \ln x + q, \tag{6.1.7}$$

where q has a value of 4.25 cal/deg·mole for the liquid alloys. This seems a fairly reasonable value for the (unknown) entropy of fusion of graphite. However, we cannot conclude with certainty from the experimental data that this is really the fusion entropy of graphite, since the solutions may show deviations from regularity.

From (6.1.5), (6.1.6), (6.1.7) and Table 10 follows for the activity of carbon in liquid iron

$$\log a_C = \log x_C + \frac{1400}{T} - 0.928, \tag{6.1.8}$$

provided that the carbon concentration is not greater than 0.1 weight% (x_C not greater than 0.00465). If the carbon contents lie between 0.1 and 0.3 weight % then the application of this equation introduces only relatively small errors.

It will be noted in Table 10 that the entropies of solution in liquid Fe-C alloys are virtually equal to those in the austenitic alloys. Taking into consideration the statistical interpretation of entropy (Chapter 1), this probably means that the number of sites in which carbon atoms can be accommodated is about the same in the two modifications. On the other hand, the entropy of solution of carbon in body-centred cubic iron (ferrite or α iron) according

---

[1]) Regular solutions is the name given to those solutions which have the same entropy of mixing as ideal solutions, but of which the energy of mixing is not zero.

to Smith (l.c.) is about 2 cal/deg·mole greater than in γ iron. This can be partly explained by the well-known facts that the carbon atoms in the two modifications are present in the octahedral interstices [1,2]) and that the number of these interstices is three times as large in α iron as in γ iron.

The data on the solution entropies of carbon in α and δ iron are less reliable than those in γ and liquid iron. The same is true for the heat of solution which, according to Smith (l.c.), is about 20,000 cal/mole for carbon in ferrite. Ferrite-austenite-liquid iron, in that sequence, show a clear decrease in the heat of solution of graphite, i.e. the heat which must be supplied for the isothermal solution of 1 mole C:

$$\alpha \quad 20{,}000 \text{ cal/mole,}$$
$$\gamma \quad 10{,}800 \text{ cal/mole,}$$
$$\text{liq} \quad 6{,}400 \text{ cal/mole.}$$

This order of succession is easily understood if it is remembered that when carbon is dissolved in solid iron, extra energy is required for the elastic straining of the lattice. This extra energy is considerably greater in ferrite than in austenite, since the octahedral interstices in ferrite are much smaller.

> A smaller value is obtained for the heat of solution of graphite in ferrite if we make use of the formation enthalpy of cementite Fe_3C, which is 6400 cal/mole at 700 °K [3]), and the enthalpy of solution of this compound in ferrite, which, according to measurements of internal friction, is 9700 cal/mole [4]). The enthalpy of solution of graphite, according to these data is $6400 + 9700 = 16{,}100$ cal/mole at 700 °K.

6.2. Activity of oxygen in liquid iron

As we have already mentioned in Section 4.5, oxygen exhibits only an extremely small solubility in the modifications of solid iron, while it has a definite (though small) solubility in liquid iron.

The equilibrium between liquid iron and liquid oxide with the approximate composition FeO, has been studied by various workers. According to Chip-

---

[1]) N. J. PETCH, J. Iron Steel Inst. **145**, 111P (1942)
[2]) G. K. WILLIAMSON and R. E. SMALLMAN, Acta Cryst. **6**, 361 (1953)
[3]) O. KUBASCHEWSKI and J. A. CATTERALL, *Thermochemical Data of Alloys*, Pergamon Press, London, 1956
[4]) L. J. DIJKSTRA, Trans. AIME **185**, 252 (1949); C. A. WERT, Trans. AIME **188**, 1242 (1950)

man and his co-workers[1]) the solubility of oxygen in liquid iron in equilibrium with FeO is given by

$$\log [\% \ O]_{max} = -\frac{6320}{T} + 2.734, \qquad (6.2.1)$$

where $[\% \ O]_{max}$ indicates the oxygen content in weight percent. The subscript "max" has been added to indicate that the equation gives the maximum quantity of oxygen which the metal can contain in homogeneous solution. In more recent investigations, Fischer and vom Ende [2]) and Gokcen [3]) find solubilities which agree well with those given by equation (6.2.1).

In the older literature one finds discussions over the question whether in cases such as that discussed here, one is dealing with solutions of oxygen or with solutions of oxide in the metal. The analogous problem was discussed for solutions of nitrogen (nitride) and hydrogen (hydride) in metals. From a thermodynamic standpoint these are only seeming problems. If a metal oxide is in equilibrium with both the corresponding metal and with gaseous oxygen, then the metal and oxygen are also in equilibrium with one another (Fig. 25). One may thus equally well speak of a solution of oxygen as of a solution of oxide. From the atomic point of view it is rather more satisfactory to speak of dissolved oxygen, since the oxygen in homogeneous solid and liquid solutions is, in general, not present in the form of oxide molecules but in that of atoms or ions.

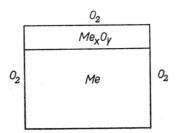

Fig. 25. If a metal oxide is in equilibrium with both the corresponding metal and gaseous oxygen, then the metal and oxygen are in equilibrium too.

[1] J. CHIPMAN and K. L. FETTERS, Trans. Amer. Soc. Metals **29**, 953 (1941); C. R. TAYLOR and J. CHIPMAN, Trans. AIME **154**, 228 (1943); J. CHIPMAN in *Basic Open Hearth Steelmaking*, published by AIME, New York, 1944, p. 472
[2] W. A. FISCHER and H. VOM ENDE, Arch. Eisenhüttenwes. **23**, 21 (1952); see also H. SCHENCK and W. PFAFF, Arch. Eisenhüttenwes. **32**, 741 (1961)
[3] N. A. GOKCEN, Trans. AIME **206**, 1558 (1956); see also E. S. TANKINS, N. A. GOKCEN and G. R. BELTON, Trans. AIME **230**, 820 (1964)

At small concentrations there is the following relationship between the mole fraction of FeO and the weight percentage of oxygen:

$$[x_{FeO}] = 0.0349 \, [\% \, O]. \tag{6.2.2}$$

Because the solubility is so small one may therefore write for (6.2.1)

$$\log [x_{FeO}]_{max} = -\frac{6320}{T} + 1.277. \tag{6.2.3}$$

According to equations (6.2.1) and (6.2.3) the solubility of oxygen in liquid iron at 1600 °C is only 0.229 weight % ($x_{FeO} = 0.00800$). It is so small that one will certainly introduce no great errors if it is assumed that for all solutions of oxygen in iron the activity is proportional to the mole fraction. For the saturated solution (in equilibrium with "FeO") we may write to a good approximation

$$[a_{FeO}]_{max} = 1 = k \, [x_{FeO}]_{max}.$$

For unsaturated solutions the following is thus valid

$$a_{FeO} = kx_{FeO} = \frac{x_{FeO}}{[x_{FeO}]_{max}}, \tag{6.2.4}$$

or, substituting in (6.2.3):

$$\log a_{FeO} = \log x_{FeO} + \frac{6320}{T} - 1.277. \tag{6.2.5}$$

6.3. The reaction between carbon and oxygen in liquid iron

If liquid iron contains both oxygen and carbon in solution, then the gas phase which is in equilibrium with the solution will consist of carbon monoxide and carbon dioxide. The reactions which lead to the establishment of equilibrium, can be written as follows:

$$[FeO] + [C] \rightleftharpoons [Fe] + CO, \tag{6.3.1}$$

$$2[FeO] + [C] \rightleftharpoons 2[Fe] + CO_2, \tag{6.3.2}$$

$$[FeO] + CO \rightleftharpoons [Fe] + CO_2, \tag{6.3.3}$$

$$[C] \quad + CO_2 \rightleftharpoons 2 \, CO, \tag{6.3.4}$$

where the square brackets indicate the components of the liquid metallic phase.

The four equations are not independent. Since (6.3.3) is equal to the difference between (6.3.2) and (6.3.1) and (6.3.4) to twice (6.3.1) minus (6.3.2), two of the four equations are sufficient to describe the equilibrium between metal and gas phase. Furthermore, in all technically important cases, the gas consists chiefly, and usually almost entirely, of CO, so that the first of the four reactions is by far the most important. It is of fundamental importance in steelmaking and in arc-welding of steel. Its reaction constant is given by

$$K_1 = \frac{p_{CO} \cdot a_{Fe}}{a_{FeO} \cdot a_C}.$$

Since we shall restrict ourselves to a consideration of dilute solutions of C and FeO in liquid iron, we can insert the value 1 for a_{Fe} without introducing appreciable errors:

$$K_1 = \frac{p_{CO}}{a_{FeO} \cdot a_C}. \tag{6.3.5}$$

Furthermore, in this case, we can insert for a_C and a_{FeO} the values given by equations (6.1.8) and (6.2.5), provided that it may be assumed that the mutual interaction of carbon and oxygen in the dilute solutions is so small that their activities are not noticeably changed by it.

The value of K_1 at an arbitrary temperature is equal to that of the constant for the reaction

$$FeO(l) + C(s) \rightleftharpoons Fe(l) + CO, \tag{6.3.6}$$

in which "FeO", C and Fe occur in the pure state (the symbols l and s indicate the liquid and solid states, i.e. the states which were chosen earlier as reference states, the states of unit activity). We can calculate approximately the constant for (6.3.6), i.e. K_1, by starting with the thoroughly investigated reaction (4.6.5). If we neglect the deviations in the stoichiometry of FeO, we can write equation (4.6.5) simply as

$$FeO(s) + CO \rightleftharpoons Fe(s) + CO_2. \tag{6.3.7}$$

This equilibrium has been studied, for a number of temperatures below the melting points of "FeO" and Fe, by Darken and Gurry [1]. In the region

[1] L. S. DARKEN and R. W. GURRY, J. Amer. Chem. Soc. 67, 1398 (1945).

$800°$-1350 °C the reaction constant of (6.3.7) is accurately given by

$$\log K_7 = \frac{850}{T} - 1.068 .$$ (6.3.8)

This is demonstrated by Fig. 26. However, we are more interested in the constant for the reaction

$$FeO(l) + CO \rightleftharpoons Fe(l) + CO_2,$$ (6.3.9)

in which FeO and Fe both participate in the liquid state. We can calculate this constant from K_7 (equation 6.3.8) by making use of our knowledge of the heats of fusion and entropies of fusion of FeO and Fe.

The heat of fusion of iron is 3300 cal/mole [1]), the melting point 1812 °K, so that the entropy of fusion is 1.82 cal/deg·mole. "FeO" with the most iron-rich composition has, according to Darken and Gurry [2]), a heat of fusion of 7230 cal/mole and an entropy of fusion of 4.40 cal/deg·mole. The transition from reaction (6.3.7) to reaction (6.3.9) is thus accompanied by a change in the reaction enthalpy of $3300 - 7230 = -3930$ cal and by

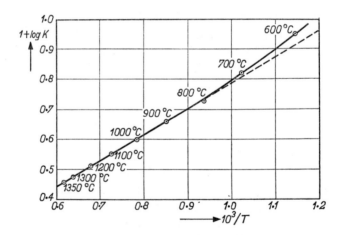

Fig. 26. Logarithm of the equilibrium constant of the reaction $Fe_{0.95}O + CO \rightleftharpoons 0.95 Fe + CO_2$ as a function of the reciprocal of the absolute temperature. The results obtained by Darken and Gurry in the region of $800°$ to 1350 °C are accurately represented by a straight line. The crystallographic transformation $\alpha_{Fe} \rightarrow \gamma_{Fe}$ does not manifest itself on this line. This is why no transition heats or transition entropies have been included in our calculation of equilibrium (6.3.9).

[1]) A. FERRIER, Compt. rend. 254, 104 (1962)
[2]) L. S. DARKEN and R. W. GURRY, J. Amer. Chem. Soc. 68, 798 (1946)

a change in the reaction entropy of $1.82 - 4.40 = -2.58$ cal/deg. Using these values and the well-known equation

$$\log K = -\frac{\Delta H^0}{4.575\ T} + \frac{\Delta S^0}{4.575},$$

(6.3.8), corresponding to reaction (6.3.7), becomes

$$\log K_9 = \frac{1710}{T} - 1.632,\qquad\qquad (6.3.10)$$

which is valid for reaction (6.3.9).

We have finally reached the point where we can find the required equilibrium constant for reaction (6.3.6) by also making use of the well-known Boudouard equilibrium, given by equation (6.1.4). Addition of the two equations

$$FeO(l) + CO \rightleftharpoons Fe(l) + CO_2 \qquad\qquad (6.3.9)$$

$$C(s) + CO_2 \rightleftharpoons 2\ CO \qquad\qquad\qquad (6.1.1)$$

gives $FeO(l) + C(s) \rightleftharpoons Fe(l) + CO, \qquad\qquad (6.3.6)$

so that $\log K_1$, corresponding to (6.3.6) and also to (6.3.1), is given by the sum of (6.3.10) and (6.1.4):

$$\log K_1 = -\frac{6725}{T} + 7.203.\qquad\qquad (6.3.11)$$

Using this equation and equations (6.3.5), (6.1.8) and (6.2.5), we find for reaction (6.3.1):

$$\log \frac{p_{CO}}{[x_{FeO}]\,[x_C]} = \frac{995}{T} + 4.998.\qquad\qquad (6.3.12)$$

For small concentrations of FeO and C in the molten iron the following equations are valid

$$[x_{FeO}] = 0.0349\ [\%\ O],$$
$$[x_C]\ \ \ = 0.0465\ [\%\ C].\qquad\qquad (6.3.13)$$

Substituting these in (6.3.12) we can write

$$\log \frac{p_{CO}}{[\%\ O]\,[\%\ C]} = \frac{995}{T} + 2.208.\qquad\qquad (6.3.14)$$

6.4. Activity and reference state

In Section 5.5 the activity of a substance i in solution was defined in such a way that for the pure state of the substance it has the value 1, i.e. the same value as the mole fraction. We expressed this by saying that the pure substance i had been chosen as reference state.

In our discussions of solutions of carbon and oxygen in iron (Sections 6.1 and 6.2) the reference states originally chosen were solid carbon in the form of graphite and liquid "FeO". Without explicitly stating this, we have changed over to other reference states in the above. For example, equation (6.3.12) is based on activities which do *not* tend to equality with the mole fractions as the pure states of C and "FeO" are approached, but on activities which are equal to the mole fractions in an infinitely dilute solution. Equation (6.3.14) is based on activities which are equal to the weight percentages in an infinitely dilute solution. To avoid confusion and difficulties, we shall, in this section, discuss the question of the choice of various reference states. The discussion of the reaction between carbon and oxygen in liquid iron will not be continued until the next section.

A more general definition of the concept of activity than that given in Section 5.5, is as follows. The activity a_i of a component i in a homogeneous solution is given by the equilibrium pressure p_i of this substance above the solution, divided by the equilibrium pressure p_i^* above some solution which has been chosen as reference state:

$$a_i = \frac{p_i}{p_i^*}. \tag{6.4.1}$$

The limiting cases are that the reference solution is extremely dilute or consists of the pure substance i. Equation (6.4.1) shows that the activity of component i in the reference state is always equal to 1, whatever this state may be.

It is frequently convenient to define the activity of the dissolved substance in such a way that the activity and the mole fraction are identical at very great dilution. Instead of the condition

$$a_i = x_i \quad \text{for} \quad x_i \to 1$$

as in Section 5.5, we now have the condition

$$a_i' = x_i \quad \text{for} \quad x_i \to 0. \tag{6.4.2}$$

It is obvious that these two activities have different values.

For very small values of x_i, the law of Henry is valid

$$p_i = kx_i, \tag{6.4.3}$$

according to which the vapour pressure of the dissolved substance at very small concentrations is proportional to the concentration (see the dotted tangents in Fig. 20). Employing (6.4.3), the condition (6.4.2) can also be formulated as follows:

$$a_i' = \frac{p_i}{k} \quad \text{for} \quad x_i \to 0. \tag{6.4.4}$$

Comparison of (6.4.1) and (6.4.4) shows that this condition amounts to a choice of a reference state for which $p_i^* = k$. This state has no physical significance. According to equation (6.4.3), it is the hypothetical state in which the dissolved substance would be found in its pure state if Henry's law were valid from $x_i = 0$ to $x_i = 1$. In the cases illustrated by Fig. 20 it corresponds to the point of intersection of the Henry-tangent with the vertical axis $x_2 = 1$.

When studying dilute solutions, it is sometimes more convenient not to choose this hypothetical state as reference state, but to choose a solution containing 1 atom % or 1 mole % of i, for which Henry's law is still valid. In technology, this is often carried a step further by choosing a solution containing 1 weight % of i as reference state. The requirement then becomes

$$a_i'' = [\% \, i] \quad \text{for} \quad [\% \, i] \to 0. \tag{6.4.5}$$

In the concentration range in which Henry's law is applicable, the activity is thus equal to the weight percentage. This range does not, in general, extend to a concentration of 1 % of substance i. In other words, the reference state is again a hypothetical state, viz. the above-mentioned solution containing 1 weight % of i for which Henry's law is still valid. In Fig. 20 it is represented by a point on the Henry tangent. If we write Henry's law in the form

$$p_i = k'[\% \, i],$$

then the condition (6.4.5) becomes:

$$a_i'' = [\% \, i] = \frac{p_i}{k'} = \frac{p_i}{p_i^*(1 \%)} \quad \text{for} \quad [\% \, i] \to 0. \tag{6.4.6}$$

Proceeding as in Section 5.1, one can immediately deduce the chemical potential of substance i in the solution:

$$\mu_i^{\text{sol}} = \mu_i^*(1 \%) + RT \ln a_i, \tag{6.4.7}$$

where $\mu_i^*(1\%)$ is the chemical potential of substance i in the hypothetical solution containing 1% of i, which still obeys Henry's law, i.e. which still behaves as an infinitely dilute solution. Making use of equation (2.2.7) one can derive from (6.4.7) the equation

$$\Delta G^* = -RT \ln \Pi \, p_j{}^{\nu_j} \, a_k{}^{\nu_k} \qquad (6.4.8)$$

which is analogous to (5.5.3). For the sake of simplicity we have omitted the double accent on the letter a. Also in the following, we shall consistently indicate any activity simply by the letter a. The reference state to which it relates will always be obvious from the text. In contrast to ΔG^0, ΔG^* is *not* the algebraic sum exclusively of μ's of pure substances and of gases at 1 atm (which are supposed to behave as perfect gases). It is seen from (6.4.7) that ΔG^* also includes μ's of substances in 1% solutions, which are assumed to behave as infinitely dilute solutions. Thus ΔG^* cannot be derived from Table 1.

According to (6.4.6), the activities in (6.4.7) and (6.4.8) may be replaced by weight percentages provided these last are very small. If the percentages are so large that appreciable deviations from Henry's law occur, we write

$$a_i = f_i[\% \, i], \qquad (6.4.9)$$

which gives the definition of the activity coefficient f for the choice of an idealized 1% solution as reference state. It is self-evident that the activity coefficient defined by (6.4.9) has a value differing from that discussed in Section 5.5. The new activity coefficient is a measure of the deviation from Henry's law, while the first was a measure of the deviation from Raoult's law.

Finally, we ask how the activity with respect to a particular reference state can be calculated from that with respect to another reference state. Suppose, for example, that we know the activity a_1 of carbon in liquid Fe-C alloys with respect to graphite as reference state:

$$\mu = \mu^0(\text{graph.}) + RT \ln a_1. \qquad (6.4.10)$$

We require to know the activity a_2 of carbon which in infinitely dilute solution is equal to the mole fraction x. As appears from the above, the reference state in this case is given by the hypothetical state in which liquid carbon in the pure state would be found if the law of Henry were to remain applicable from $x = 0$ to $x = 1$:

$$\mu = \mu^*(\text{C hyp.}) + RT \ln a_2. \qquad (6.4.11)$$

From (6.4.10) and (6.4.11) follows

$$\mu^*(\text{C hyp.}) - \mu^0(\text{graph.}) = RT \ln \frac{a_1}{a_2}. \tag{6.4.12}$$

The left-hand side of the equation is nothing other than the change in free enthalpy which occurs by the transition of 1 mole graphite into the above-mentioned hypothetical carbon, i.e. by the transition of 1 mole carbon from one reference state to the other. The value of this is given, from (6.1.6), (6.1.7) and Table 10, by

$$\mu^* (\text{C hyp.}) - \mu^0 (\text{graph.}) = 6400 - 4.25 \, T. \tag{6.4.13}$$

From (6.4.12) and (6.4.13) follows:

$$\log a_1 = \log a_2 + \frac{1400}{T} - 0.928. \tag{6.4.14}$$

The activity a_2 can be calculated from a_1 (the activity with respect to graphite) with the help of this equation. Only for very small concentrations is a_2 equal to the mole fraction x of carbon.

If we compare (6.1.8) with (6.4.14), we see that even as early as Section 6.1 we made use of equation (6.4.14), but without realizing that it had any connection with a transition from one reference state to another. An analogous remark may be made about equation (6.2.5).

6.5. Further consideration of the reaction between carbon and oxygen in liquid iron

In Section 6.3, the equilibrium between carbon and oxygen in liquid iron was calculated by commencing with the well-known data on the equilibria between solid Fe + "FeO" and mixtures of the gases CO + CO_2. Various investigators have tried, with more or less success, to determine the equilibrium by direct experimental means [1]. This involves great difficulties because the concentrations of carbon and oxygen which were present in the liquid metal during the experiment must be determined later (after solidification). The solidification of the metal is accompanied by evolution of gas and the consequent loss of carbon and oxygen.

---

[1] H. C. VACHER and E. H. HAMILTON, Trans. AIME **95**, 124 (1931); H. C. VACHER, Bur. Stand. J. Res. **11**, 541 (1933); S. MARSHALL and J. CHIPMAN, Trans. Amer. Soc. Metals **30**, 695 (1942).

The equilibrium between carbon and oxygen in liquid iron can also be calculated directly from the equilibria (6.3.3) and (6.3.4), experimentally determined by Gokcen [1]) and Richardson et al. [2]). These reactions can also be written in the form

$$[O] + CO \rightleftharpoons CO_2, \qquad (6.5.1)$$

$$[C] + CO_2 \rightleftharpoons 2\ CO. \qquad (6.5.2)$$

Adding these, we obtain the required equilibrium

$$[O] + [C] \rightleftharpoons CO, \qquad (6.5.3)$$

which is similar to (6.3.1).

For the first of these three reactions, the experiments of Gokcen (l.c.) provide us with the equation

$$\log \frac{p_{CO_2}}{[\%\ O] \cdot p_{CO}} = \frac{8088}{T} - 4.438 . \qquad (6.5.4)$$

The most reliable basis for equation (6.5.4) is not formed by the direct measurements by Gokcen of the equilibrium (6.5.1), but by his measurements of the corresponding reaction with hydrogen

$$[O] + H_2 \rightleftharpoons H_2O. \qquad (6.5.5)$$

From these results equilibrium (6.5.1) could be calculated by making use of the water-gas equilibrium

$$H_2 + CO_2 \rightleftharpoons H_2O + CO, \qquad (6.5.6)$$

which is well-known from spectroscopic data. By subtracting (6.5.6) from (6.5.5) one obtains (6.5.1). The result given by (6.5.4) corresponds satisfactorily with the direct measurements of equilibrium (6.5.1).

The equilibrium constant of reaction (6.5.2) follows from (6.1.4), (6.1.8) and (6.3.13):

$$\log \frac{p_{CO}^2}{[\%\ C] \cdot p_{CO_2}} = -\frac{7035}{T} + 6.574 . \qquad (6.5.7)$$

By adding (6.5.4) and (6.5.7) we obtain the required constant for reaction (6.5.3):

$$\log \frac{p_{CO}}{[\%\ O] \cdot [\%\ C]} = \frac{1053}{T} + 2.136 . \qquad (6.5.8)$$

[1]) N. A. GOKCEN, Trans. AIME 206, 1558 (1956).
[2]) F. D. RICHARDSON and W. E. DENNIS, Trans. Faraday Soc. 49, 171 (1953).

In a manner similar to that followed in this section, Fuwa and Chipman [1]) find:

$$\log \frac{p_{CO}}{[\% O] \cdot [\% C]} = \frac{1168}{T} + 2.07 . \qquad (6.5.9)$$

There is satisfactory agreement between equations (6.3.14), (6.5.8) and (6.5.9). At a temperature of 1800 °K and a CO pressure of 1 atm, they provide the following values for the product $[\% O] \cdot [\% C]$: 1.74×10^{-3}, 1.90×10^{-3} and 1.91×10^{-3}. The differences between these figures fall within the limits of experimental accuracy, so that there is no reason to prefer any one of the three equations above the others. As a compromise we suggest the following:

$$\log \frac{p_{CO}}{[\% O] \cdot [\% C]} = \frac{1070}{T} + 2.135 , \qquad (6.5.10)$$

which for 1800 °K and 1 atm CO, gives a figure of 1.87×10^{-3} for the product $[\% O] \cdot [\% C]$.

As follows from the above, formula (6.5.10) is, strictly speaking, only valid for very dilute solutions of carbon and oxygen in liquid iron. At increasing concentrations, even in the binary alloys Fe-C and Fe-O, appreciable departures from Henry's law occur. Activity and weight percentage can then no longer be equated. In the formulae we must replace $[\% C]$ and $[\% O]$ by $a_C = f_C [\% C]$ and $a_O = f_O [\% O]$ (cf. Section 6.4).

Fig. 24, which has already been discussed, shows the deviations from Henry's law for carbon in liquid iron. The activity coefficient for carbon increases with increasing concentration. In contrast to this, the activity coefficient for oxygen in liquid iron, according to investigations by Floridis and Chipman [2]), decreases with increasing concentration.

For $\log f_O$ in the equation

$$\log a_O = \log f_O + \log [\% O]$$

these investigators give:

$$\log f_O = - 0.20 [\% O]. \qquad (6.5.11)$$

As it must do, f_O approaches 1 as $[\% O] \rightarrow 0$.

---

[1]) T. Fuwa and J. Chipman, Trans. AIME **218**, 887 (1960).
[2]) T. P. Floridis and J. Chipman, Trans. AIME **212**, 549 (1958).

Besides the above-mentioned deviations from Henry's law in the binary alloys Fe-C and Fe-O, in the ternary alloys Fe-C-O one must also take into account the interaction between carbon and oxygen. In those cases where this has been investigated, it could be shown that carbon and oxygen dissolved in *solid* metals behave as positive and negative ions: In an electric field carbon moves through solid iron to the negative electrode [1,2]), while oxygen in solid zirconium moves towards the positive electrode [3]). Little is known of the structure of the corresponding liquid solutions, but the above-mentioned observations make it seem likely that carbon and oxygen are present in the form of positive and negative ions in liquid iron. This would result in a mutual reduction of the two activity coefficients.

In agreement with this prediction, Fuwa and Chipman [4]) find that the activity coefficient of oxygen in liquid iron is lowered by the presence of comparatively large concentrations of carbon. They express this interaction in the equation

$$\log f_O^{(C)} = -0.13 \, [\% \, C]. \qquad (6.5.12)$$

A corresponding lowering of the activity coefficient of carbon in liquid iron (as a result of the presence of dissolved oxygen) is difficult to observe because of the small solubility of oxygen as compared with that of carbon.

Fig. 24 and equations (6.5.11) and (6.5.12) show that the activity coefficients of carbon and oxygen in liquid iron change in opposite directions when their concentrations increase. The most important result of this is that the product $[\% \, C] \cdot [\% \, O]$ at constant CO pressure and constant temperature only exhibits a relatively small dependence on the C and O concentrations. For all concentrations which are of importance in steelmaking and electric arc-welding of steel, it is justifiable in practical discussions to accord this product a constant value at constant temperature. It should be realized in this connection that a large carbon concentration corresponds to only a very small oxygen concentration (for a C concentration of 2%, an O concentration of only about 0.001% at a CO pressure of 1 atm). The uncertainty in the analytical determination of such a small oxygen content is proportionally much larger than the deviation of the product $[\% \, C] \cdot [\% \, O]$ from a mean value. The latter deviations can thus be neglected. According to equation (6.5.10) the dependence on temperature is also relatively small.

---

[1]) W. SEITH and O. KUBASCHEWSKI, Z. Elektrochem. **41**, 551 (1935).
[2]) P. DAYAL and L. S. DARKEN, Trans. AIME **188**, 1156 (1950).
[3]) J. H. DE BOER and J. D. FAST, Receuil Trav. Chim. Pays-Bas **59**, 161 (1940).
[4]) T. FUWA and J. CHIPMAN, Trans. AIME **218**, 887 (1960).

Table 11 gives the calculated values of the product $[\% \, C] \cdot [\% \, O]$ for various temperatures and carbon concentrations at an equilibrium pressure of 1 atm.

TABLE 11

VALUES OF THE PRODUCT $[\% \, C] \cdot [\% \, O] \cdot 10^3$ IN LIQUID IRON AT A PRESSURE OF 1 ATM (CO + CO₂) ACCORDING TO FUWA AND CHIPMAN (L.C.)

$[\% \, C]$	1500 °C	1600 °C	1700 °C	1800 °C	1900 °C
0.02-0.20	1.86	2.00	2.18	2.32	2.45
0.5	1.77	1.90	2.08	2.20	2.35
1.0	1.68	1.81	1.96	2.08	2.25
2.0	1.55	1.70	1.84	1.95	2.10

The values of the product $[\% \, C] \cdot [\% \, O]$ in the table do not relate to a CO pressure but to a (CO + CO₂) pressure of 1 atm. The gas phase in equilibrium with the solutions in question contains not only CO but also CO_2. The CO_2 pressure corresponding to any particular content of carbon and oxygen can easily be calculated from the equilibria

$$2 \, [O] + 2 \, [C] \rightleftharpoons 2 \, CO$$

$$[C] + CO_2 \rightleftharpoons 2 \, CO$$

subtracting $\qquad 2 \, [O] + [C] \rightleftharpoons CO_2.$ \hfill (6.5.13)

The constant for reaction (6.5.13) can be found by subtracting (6.5.7) from twice (6.5.10):

$$\log \frac{p_{CO_2}}{[\% \, O]^2 \cdot [\% \, C]} = \frac{9175}{T} - 2.304 \, . \qquad (6.5.14)$$

We shall demonstrate the application of the formulae discussed by calculating the equilibrium pressures of CO and CO_2 above liquid iron at 1600 °C, containing 0.20% C and 0.01% O. This is done with the help of (6.5.10) and (6.5.14) after $[\% \, O]$ in these equations has been replaced by $f_O^{(C)} [\% \, O]$. According to Fig. 24 it is unnecessary to apply a correction to the carbon content, $[\% \, C]$. Using equations (6.5.10), (6.5.14) and (6.5.12) we find: $p_{CO} = 0.96$ atm and $p_{CO_2} = 0.007$ atm. The latter pressure can, of course, be calculated from the former with the help of equation (6.5.7).

We have already mentioned that in practical discussions $[\% \, O] \cdot [\% \, C]$ is treated as being independent of the oxygen and carbon contents. Calculations are often made even more roughly by considering this product as also independent of the temperature and, furthermore, neglecting the presence of CO_2 in the gas.

6.6. Evolution of CO when steel solidifies

Great technical importance can be attached to the formation of CO when steel solidifies. This subject is directly connected to the above and will therefore be treated in this section, although the questions to be discussed lie, for the greater part, outside the domain of thermodynamics.

The solubility of carbon and oxygen in solid iron is much smaller than in liquid iron. In the neighbourhood of the melting point of pure iron, the binary equilibrium diagrams Fe-C and Fe-O are both of the type illustrated in Fig. 27a. Let us suppose that liquid iron, containing carbon and oxygen, is in equilibrium with CO in the gaseous phase (for the sake of simplicity we shall ignore the small percentage of CO_2 in the gas). If the metal is allowed to solidify slowly then, at first, solid metal will be formed, which is much purer than the liquid. In other words, the growing iron crystals expel carbon and oxygen, so that the concentrations of these elements in the liquid metal increase. The liquid becomes supersaturated with C and O with respect to the gas phase and there will be a tendency to form gas.

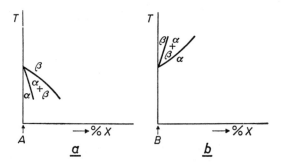

Fig. 27. (a) The melting point or a crystallographic transition point of a component (A) is shifted to lower temperatures by the addition of a second component (X) when the solubility of X in the α phase, stable at low temperatures, is smaller than that in the β phase, stable at high temperatures. (b) In the reverse case (system B-X), the melting point or transition point is shifted to higher temperatures.

Case (a) applies to the melting points in the Fe-C and Fe-O systems.

FORMATION OF GAS BUBBLES

The gas can be produced in two ways: (1) The carbon and oxygen atoms can reach the outer surface by either diffusion or convection and react there to form CO molecules. (2) Gas bubbles may form in the liquid metal. In the latter case, new interfaces are formed even at great distances from the outside surface and at these new surfaces CO molecules can form from the dissolved C and O atoms. In order to do this, the atoms only have to travel a relatively short distance by diffusion.

The formation of gas bubbles during solidification is one of the most notorious sources of difficulties in metal technology, in particular in casting, welding and soldering many metals and alloys. We shall therefore discuss the circumstances under which bubbles can be formed.

A gas bubble in a liquid is only able to exist if its internal pressure is greater than the total external pressure, which is the sum of the gas pressure above the metal and the hydrostatic pressure of the column of liquid above the bubble. In other words: if a gas bubble is to arise somewhere in the liquid metal in question, then the concentration of carbon and oxygen must be so great at that point that the corresponding equilibrium pressure of CO (which we might call the *latent* pressure of the gas in the metal) is greater than the total external pressure. If, for instance, an extra pressure of 1000 atm of argon were applied above the metal, then gas bubbles would only be able to form with an internal pressure greater than 1000 atm, in other words, it would be virtually impossible for gas to escape as bubbles, although the argon has no influence on the thermodynamics of the metal-gas equilibria involved. Equilibrium could now only be established by formation of CO molecules at the outer surface.

In the above, no account has been taken of the fact that the formation of a gas bubble involves a disruption of the cohesion in the liquid metal. We shall show that the formation of the new boundary surface corresponds to an extra "capillary pressure", which is given by $2\sigma/r$, where σ is the surface tension of the liquid metal and r is the radius of the gas bubble.

Suppose that there is a gas bubble with radius r already present in the liquid. If the radius increases from r to $r + dr$, the surface area will increase from $4\pi r^2$ to

$$4\pi r^2 + 8\pi r dr . \tag{6.6.1}$$

The surface tension σ of a liquid indicates the work which must be performed on the liquid in order to increase its surface area, in a reversible and isothermal manner, by one unit of area. The expansion of the gas bubble in question will, according to (6.6.1), require an amount of work

$$dW = 8\pi\sigma r dr . \tag{6.6.2}$$

We can also write another expression for this amount of work. To do this, we shall call the gas pressure in the bubble, which is in equilibrium with the various other pressures, P. Further, we indicate the gas pressure on the liquid by p_1 and the hydrostatic pressure on the bubble by p_2. The resultant force on an element of surface $d\omega$ thus amounts to:

$$K = (P - p_1 - p_2)\, d\omega.$$

If the radius of the bubble increases by dr, then this force will perform an amount of work

$$Kdr = (P - p_1 - p_2)\, d\omega dr.$$

The total work dW, which is performed during the expansion of the bubble, is found by integration over all the surface elements:

$$dW = (P - p_1 - p_2)\, 4\pi r^2 dr \,. \tag{6.6.3}$$

By equating (6.6.2) and (6.6.3) we find for the pressure in the bubble, the equation

$$P = p_1 + p_2 + \frac{2\sigma}{r} \,. \tag{6.6.4}$$

An extra pressure of $2\sigma/r$ is thus required in the bubble to establish equilibrium with the surface tension.

A growing bubble must inevitably pass through an initial stage in which it is extremely small. In this stage, according to (6.6.4), the pressure required in the bubble is very large. To what extent the liquid iron must be supersaturated with carbon and oxygen in order that bubbles shall *appear*, cannot, however, be deduced from (6.6.4), because it is impossible to apply this formula down to atomic dimensions, since the concept of surface tension is a typically macroscopic concept which loses its significance in the range of atomic dimensions. In order to indicate at least a lower limit for the required pressure P, we apply equation (6.6.4) to a bubble with a radius of 10^{-5} cm, which seems justifiable. For the surface tension of pure liquid iron at 1550 °C, Kozakevitch and Urbain [1]) give the value 1800 ergs/cm². This, however, is greatly reduced by oxygen. Since we are only interested in the order of magnitude, we assume a value of 1000 ergs/cm² for our C- and O-bearing iron. With this value and $r = 10^{-5}$ cm, $2\sigma/r$ is about 200 atm. The required latent pressure for the *formation* of bubbles may well be greater by a factor 10 or 100.

[1]) P. KOZAKEVITCH and G. URBAIN, J. Iron Steel Inst. **186**, 167 (1957) and Mém. Scient. Rev. Mét. **58**, 401, 517, 931 (1961).

The formation of gas bubbles inside the liquid metal is thus practically impossible. Bubbles are, in fact, seen to form at the hearth-metal interface and at the boundary of the solidifying metal, especially at sharp corners and edges of growing metal crystals. This has been demonstrated by many experiments, e.g. experiments by Oelsen [1]) in which liquid iron was surrounded on all sides by liquid slag. In this case, even at a C content of 2% and an O content of 0.035%, no boiling phenomena were observed, i.e. no gas bubbles formed, although these contents correspond from equations (6.5.10) and (6.5.14) at 1550 °C to a ($CO + CO_2$) pressure of about 40 atm (cf. Table 11). This latent pressure was not sufficient to cause gas bubbles to form in the liquid metal. When an iron rod was immersed in the melt supersaturated with carbon and oxygen, then violent evolution of gas occurred, which immediately ceased when the rod was withdrawn.

The influence of the iron rod may be compared with that of "boiling chips" used to avoid retardation of boiling in water and other liquids. Nevertheless, even the formation of gas bubbles on metal crystals requires an extra supersaturation. It is significantly smaller, however, than corresponds to the quotient $2\sigma/r$.

If one wishes to prevent the occurrence of gas cavities in the solidified metal, one must try to ensure that the last metal to solidify is not already enclosed on all sides by solidified metal. In that case, the gas bubbles can no longer escape and when the metal has completely solidified it will contain cavities and pores. One must try to achieve "directional solidification", i.e. solidification which begins at the bottom and slowly proceeds upwards, in such a way that nowhere is liquid metal shut in by metal which has already solidified. This method of solidification not only greatly reduces the chance of inclusion of gas bubbles, but also avoids the formation of internal shrinkage cavities and fissures.

As we have seen above, gas bubbles form at the interface between the solid and liquid metal. As long as they are small, they adhere to this boundary. Only when they exceed a certain size can they break away from the solid wall, rise to the surface and escape, under the influence of the hydrostatic pressure. If the bubbles grow more slowly or not much faster than the rate at which the interface between solid and liquid moves with the progress of the solidification, then even with directional solidification gas cavities will remain behind in the metal in the form of long gas inclusions. Their longest dimension is approximately perpendicular to the solidifying surface. This is illustrated by Fig. 28.

---

[1]) W. OELSEN, Stahl Eisen **56**, 182 (1936).

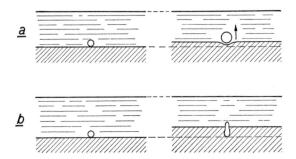

Fig. 28. Gas bubbles in solidifying steel form by preference at the boundaries between liquid and solid metal. If they grow rapidly enough, they can break free in time and, in the case of directional solidification, can escape (part *a* of the figure). If they grow too slowly, they may be enclosed by the advancing solid metal and then give rise to internal cavities (part *b* of the figure). This is facilitated if the boundary between solid and liquid does not move as a smooth plane, as has been assumed in the figure, but proceeds irregularly due to the growth of crystals with many branches (dendrites).

As a special case of this race between a growing gas bubble and the progress of an interface, it may happen that the lower part of the bubble remains behind, while at the top, where the bubble has already increased considerably in size, a portion may suddenly break free. The liquid metal, flowing in, will here cause a narrowing of the gas cavity, which in the mean time has grown further. This process may be repeated a number of times so that cavities are formed which periodically narrow and widen, and which have a very bad reputation in both steel castings and welds [1]).

PREVENTION OF GAS EVOLUTION DURING SOLIDIFICATION

It is possible to treat liquid steel so that no bubbles are formed on solidification. This goal can be achieved in various ways, e.g. by (1) chemical combination of the oxygen present in the steel, (2) treatment of the liquid metal in vacuum, (3) allowing an inert gas to bubble through the metal, (4) generating ultrasonic vibrations in the liquid metal. The first two methods are applied on a large scale in steelmaking. One of these (the vacuum treatment) is usually not primarily intended to prevent the formation of gas pockets in the cast metal, but to remove hydrogen from the metal.

The first method consists of adding elements to the liquid steel which have a greater affinity for oxygen than has the iron. The dissolved oxygen unites with these elements to form very stable oxides which react little or not at all

---

[1]) Cf. also A. HULTGREN and G. PHRAGMEN, Trans. AIME **135**, 133 (1939).

with carbon during solidification. One of the most frequently used oxygen binders is silicon which is added in the form of ferro-silicon. In agreement with Fig. 2 and Table 1, aluminium proves to be an even more effective deoxidizer than silicon. Aluminium has, in addition, a greater affinity for nitrogen which enables it, when added in sufficient quantities, to combine also with the dissolved nitrogen. It should be remembered that commercial steel, besides carbon and oxygen, always contains varying amounts of hydrogen and nitrogen, so that the escaping gas contains H_2, H_2O and N_2 (and traces of CH_4) in addition to CO and CO_2. The quantities of hydrogen and nitrogen which remain in the steel can have an adverse effect on the properties of the metal after cooling (see the following chapters). Metals such as titanium, zirconium, vanadium and niobium, like aluminium, can render both oxygen and nitrogen harmless. Steel which has been treated with silicon, aluminium or one of the other active metals is known as "killed steel", since the violent generation of gas no longer takes place on solidification. If one is still troubled by the formation of pores during the solidification of killed steel, this will be due chiefly to the liberation of hydrogen (see Section 7.4). The killing process just described has the disadvantages that it does not remove the hydrogen from the melt and that the solidified metal contains oxide and silicate inclusions.

> Steel which has not been "killed" is also produced in large quantities. During solidification a great deal of gas is liberated, mainly CO. The ingots are allowed to solidify in a non-directional manner, so that many gas bubbles are trapped in the solidifying metal. The unkilled steel is therefore of a lower quality for many purposes than killed steel. It is generally called "rimmed steel" because it has an outer skin or "rim" of relatively pure iron around a segregated core which contains the greater part of the carbon, phosphorus and sulphur. It has the advantage that the decrease in volume which occurs on solidification (solid iron has a smaller volume than molten iron) is approximately compensated by the increase in volume which accompanies the formation of blow-holes. In this way no significant loss of metal occurs, whereas in the case of killed steel the upper portion of any ingot must be sheared off due to the formation of a large shrinkage cavity or "pipe". The internal cavities remaining in rimmed steel are closed when the ingots are hot-rolled.

Increasing quantities of steel are being produced which undergo a vacuum-treatment while in the molten state [1]. One speaks of *vacuum melting* if this

[1]   See e.g. *Vacuum Metallurgy*, ed. R. F. BUNSHAH, Reinhold Publ. Corp., New York, 1958, and Trans. Vacuum Metallurgy Conference, 1960, ed. R. F. BUNSHAH, Interscience Publ., New York, 1961.

treatment starts with solid metal which is melted in vacuum and kept in vacuum until it has resolidified. In contrast, one speaks of *vacuum degassing* when the metal is only vacuum-treated when liquid, after having been melted in the conventional manner.

For vacuum melting many types of induction furnaces, arc furnaces and electron-beam furnaces are in use. It is an expensive process which is principally employed for the preparation and purification of reactive metals (titanium, zirconium, etc.) and refractory metals (molybdenum, tungsten, etc.). In steelmaking, vacuum melting can only be applied to relatively small quantities of special types of steel. On the other hand, the vacuum degassing method is much less costly, while it can be applied on any desired scale. It also lends itself for the treatment of unalloyed or low-alloy steel. By this method it is not sufficient simply to pour the steel into a ladle and then to place this ladle in a chamber which is evacuated. In that case, only the upper layer of the liquid metal would be degassed, while the ferrostatic pressure would prevent the formation of gas in the deeper layers. If the degassing is to be effective, all the steel must be able to benefit from the vacuum treatment. Of the various ways in which this can be achieved we shall discuss two which are already being used in steelmaking on a large scale. These are degassing by the Dortmund process and that by the Bochum process.

Fig. 29 represents schematically the Dortmund process [1-3]. A reservoir, fitted with a nozzle at the bottom, is connected to a pump system. When the lower end of the nozzle has been placed in the liquid steel, the reservoir is evacuated, which causes the steel to rise in the nozzle to an equilibrium height

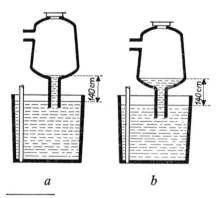

Fig. 29. Degassing of steel according to the "Vakuumheber-Verfahren" of the Dortmund-Hörder Hüttenunion.

a   *b*

[1] F. HARDERS, H. KNÜPPEL and K. BROTZMANN, Stahl Eisen **79**, 267 (1959).
[2] C. H. POTTGIESSER, H. J. DÄRMANN and A. DREVERMANN, Stahl Eisen **79**, 463, (1959)
[3] P. J. WOODING and W. SIECKMAN, Trans. Vacuum Met. Conf., 1960 (ed. R. F. Bunshah) p. 243.

of 140 cm (*a*). The reservoir is now lowered vertically until it is partly filled with steel. The shape of the reservoir is such that this steel (e.g. 10 % of the steel in the ladle) presents a large surface to the vacuum (*b*). When the reservoir is again raised the steel drops back into the ladle. If the reservoir is moved up and down frequently enough, the whole steel mass will be subjected to the vacuum treatment.

In the Bochum process [1-5]), shown schematically in Fig. 30, an initially empty ingot mould is situated in an evacuated space. A ladle containing the molten steel is sealed to the top of the vacuum vessel. When the stopper rod *S* is raised, the molten metal comes in contact with an aluminium disc which closes an opening in the vacuum vessel. The disc melts and the metal streams through the vacuum into the mould. The steel jet divides inside the vessel into a large number of droplets, which greatly helps the degassing. The gas liberated, as in the other degassing process, is pumped off continuously. Compared with the Dortmund process (Fig. 29), the Bochum process (Fig. 30) has the advantage that, after degassing, the metal does not come into contact with the air again while it is in the molten state, so that reoxidation and absorption of hydrogen by reaction with water vapour is avoided. On the other hand, the degree of degassing can be better controlled in the former process.

Due to the low pressure which is reached during vacuum degassing, the deoxidation of the steel is strongly promoted. If the constant for the reaction

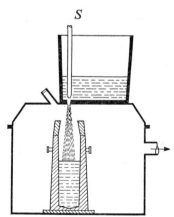

S

Fig. 30. Degassing of steel according to the "Vakuumgiessverfahren".

[1]) A. Tix, Stahl Eisen **76**, 61 (1956).
[2]) H. Hoff, H. J. Kopineck and G. Opfer, Stahl Eisen **79**, 408 (1959).
[3]) A. Tix, G. Bandel, W. Coupette and A. Sickbert, Stahl Eisen **79**, 472 (1959).
[4]) G. E. Danner and G. Taylor, Trans. Vacuum Met. Conf., 1960 (ed. R. F. Bunshah), p. 225.
[5]) L. Colombier, Revue Métall. **58**, 151 (1961).

$$[C] + [O] \rightleftharpoons CO$$

at a particular temperature has the value K, then according to (6.5.10) the following is valid for the equilibrium state

$$[\% \, C] \cdot [\% \, O] = K^{-1} p_{CO} \, .$$

It is therefore desirable to keep the partial pressure of CO above the liquid metal as low as possible. This is also true even when the available time is too short for complete establishment of equilibrium [1]).

In addition to the advantage that no cavities are formed during solidification, vacuum degassing has the advantage that the solidified steel contains less nitrogen and hydrogen than would be the case without the vacuum treatment. Furthermore, the content of oxide inclusions and silicate inclusions is smaller than in steel which has been killed in the conventional manner. As we have already mentioned, for many types of steel the great reduction of the hydrogen content is of primary importance (cf. Sections 7.4 to 7.7).

Molten steel can also be degassed in the ladle by placing this in a chamber which is then evacuated, while at the same time helium or argon is allowed to bubble through the liquid metal [2]). The gas bubbles, in which the partial pressures of CO and H_2 are initially zero, present interfaces at which gases can be formed at every level in the liquid. Instead of introducing interfaces to the metal at the bottom of the ladle in this way, one can also reverse the process and bring the metal to the surface. This can be done by generating turbulent motion in the metal by means of induction currents [3]). In vacuum *melting* of metals by means of induction heating considerable experience has already been obtained on the promotion of degassing by a circulating motion ("induction stirring") of the liquid metal.

Another method of degassing liquid metals is to generate ultrasonic vibrations in the metal. In this way local cavitation can be produced, which can serve as nuclei for gas bubbles. This method, however, does not lend itself to the degassing of large quantities of steel.

[1]) See also: H. J. KOPINECK, E. SCHULTE and A. SICKBERT, Stahl Eisen **82**, 846 (1962) and A. MUND, Stahl Eisen **82**, 1485 (1962). In these articles many other articles on vacuum degassing are cited.

[2]) C. W. FINKL, Trans. Vac. Met. Conf., 1959, Univ. Press, New York, 1960, p. 93; Metal Progress **76**, 111 (Sept. 1959).

[3]) TH. E. PERRY, Metal Progress **84**, 88 (Aug. 1963).

CHAPTER 7

SOLUTIONS OF GASES IN METALS - I

7.1. Introduction

In the preceding chapter, the liberation of CO by liquid steel according to the equation

$$[C] + [O] \rightleftharpoons CO$$

was discussed at length. This evolution of gas also takes place when heating *solid* iron which contains carbon and oxygen. A similar phenomenon is observed for various other metals. For example, it is a well-known matter of experience that commercial nickel liberates a great deal of CO when it is heated. The quantities of carbon and oxygen corresponding to the quantity of CO liberated by the solid metal are usually not wholly present in solution, but for the larger part in the form of carbide and oxide particles. At the surface of the metal CO molecules are formed from carbon and oxygen atoms which are continually replenished by diffusion from the interior of the metal. The resultant impoverishment of the solid solution of carbon and oxygen causes the carbide and oxide particles to pass gradually into solution.

If the above degassing equation is read in the reverse direction, it describes the solution of CO in a metal:

$$CO \rightleftharpoons [C] + [O].$$

The equation shows that CO, when it dissolves in a metal, does not do this as such, but split into atoms (or ions). We can imagine that CO molecules from the gas phase are adsorbed at the surface of the metal, where some of them dissociate into atoms:

$$CO_{gas} \rightleftharpoons CO_{ads},$$

$$CO_{ads} \rightleftharpoons C_{ads} + O_{ads}.$$

Adsorbed atoms subsequently diffuse into the interior of the metal until equilibrium is attained between adsorbed and dissolved atoms:

$$C_{ads} \rightleftharpoons [C],$$

$$O_{ads} \rightleftharpoons [O].$$

A statement similar to that made above for CO is valid for the solution of other diatomic gases in metals. If they do dissolve in metals, they do not do this as such, but split into atoms (or ions). The equations below demonstrate this analogy for the solution of CO, H_2, N_2 and O_2:

$$CO \rightleftharpoons [C] + [O], \qquad (7.1.1)$$

$$H_2 \rightleftharpoons [H] + [H], \qquad (7.1.2)$$

$$N_2 \rightleftharpoons [N] + [N], \qquad (7.1.3)$$

$$O_2 \rightleftharpoons [O] + [O]. \qquad (7.1.4)$$

7.2. Dependence of solubility on pressure

As an example of the solution of a homonuclear diatomic gas in a metal, we shall consider equation (7.1.2). The reaction constant of this equation is given by

$$K' = \frac{a_H^2}{p_{H_2}}. \qquad (7.2.1)$$

For small concentrations of the dissolved hydrogen, its activity is proportional to its concentration (cf. Sections 5.5 and 6.1), so that we may write for (7.2.1) in that case:

$$K'' = \frac{[\% \, H]^2}{p_{H_2}}. \qquad (7.2.2)$$

According to this equation, the quantity of hydrogen which dissolves in a metal is proportional to the square root of the H_2 pressure outside the metal:

$$[\% \, H] = K \sqrt{p_{H_2}}. \qquad (7.2.3)$$

In agreement with this equation, the solubility of homonuclear diatomic gases (hydrogen, nitrogen and oxygen) in metals is always found experimentally to be proportional to the square root of the pressure of the gas in question, as long as the concentration of the solution is small.

Historically, the road indicated was followed in the opposite direction. As early as the last century Hoitsema and Bakhuis Roozeboom found that the quantity of hydrogen which dissolves in palladium at constant temperature is proportional to the square root of the pressure of the gaseous hydrogen, as long as the solutions are very dilute [1]. From this they came

---

[1] C. HOITSEMA, Z. phys. Chem. **17**, 1 (1895).

to the correct conclusion, that the hydrogen is present in atomic form in these solutions. For many other metal-gas systems the analogous $\sqrt{p}$ law was subsequently found by Sieverts and his co-workers [1]), e.g. for the silver-oxygen system already at the beginning of this century [2]). After this other investigators have repeatedly confirmed that hydrogen, nitrogen and oxygen dissolve in atomic form in both liquid and solid metals. For the monatomic, noble gases (helium, neon, argon, etc.) one would expect a solubility proportional to the pressure. It turns out, however, that there is no metal in which they will dissolve to a measurable degree.

7.3. Dependence of solubility on temperature

Besides the dependence of the solubility on the pressure, its dependence on *temperature* is also of great importance. We consider once more the solubility of hydrogen in a metal and write the equation of solution in the form

$$\tfrac{1}{2} H_2 \rightleftharpoons [H]. \tag{7.3.1}$$

According to Sections 2.2 and 2.3, we have for this equilibrium

$$\mu_H = \tfrac{1}{2}\,\mu_{H_2} = \tfrac{1}{2}\,(\mu^0_{H_2} + RT \ln p_{H_2}), \tag{7.3.2}$$

where μ_H is the partial molar free enthalphy (chemical potential) of the hydrogen in the metal, μ_{H_2} the molar free enthalpy of the gaseous hydrogen at the pressure prevailing and $\mu^0_{H_2}$ the molar free enthalpy of (perfect) gaseous hydrogen at 1 atm pressure. Strictly speaking, equation (7.3.2) is only valid for pressures at which the gas may be regarded as perfect. In fact, in nearly all the problems which interest us this is practically true. We can also write equation (7.3.2) in the form

$$\Delta\mu_H = \mu_H - \tfrac{1}{2}\mu^0_{H_2} = \tfrac{1}{2}\,RT \ln p_{H_2}, \tag{7.3.3}$$

where $\Delta\mu_H$ is the partial molar free enthalpy of the hydrogen in the metal with respect to that of gaseous hydrogen at 1 atm as zero level. This *relative* partial molar free enthalpy, in agreement with Section 6.1, is also given by

$$\Delta\mu_H = \Delta\bar{h}_H - T\Delta\bar{s}_H, \tag{7.3.4}$$

[1]) A. SIEVERTS, Z. Metallkunde **21**, 37 (1929).
[2]) A. SIEVERTS and J. HAGENACKER, Z. phys. Chem. **68**, 115 (1909).

where $\Delta \bar{h}_{\mathrm{H}}$ and $\Delta \bar{s}_{\mathrm{H}}$ are the relative partial molar enthalpy and the relative partial molar entropy of hydrogen in the metal.

   The best method, fundamentally, by which to determine the relative partial molar quantities is the following. One first determines the values of $\Delta \mu$ for various concentrations and temperatures from measurements of pressure according to equation (7.3.3). Then $\Delta \mu$ is plotted against temperature for a number of *constant* concentrations. The slope of any of these curves gives, according to equation (7.3.4), the relative partial molar entropy $\Delta \bar{s}$ of the dissolved gas at the concentration in question. Having found values of $\Delta \mu$ and $\Delta \bar{s}$ these can be used with the help of equation (7.3.4) to find values of $\Delta \bar{h}$ for various temperatures and concentrations. In this way, for example, the relative partial molar quantities have been determined for hydrogen dissolved in vanadium [1]), niobium [2]) and tantalum [3,4]) (see Chapter 8).

> From the relative partial molar quantities of the hydrogen, those of the metal in the solution can be calculated. It is also possible to calculate the relative *integral* molar quantities from the relative *partial* molar quantities. (The relative integral molar enthalpy of any solution is the amount of heat which must be supplied during the isothermal formation of a mole of that solution from the pure components.) These relatively simple calculations are not dealt with in this book because they are of little importance in the metal-gas systems to be discussed.

   If we are considering only dilute solutions of gases in metals, we can make use of equation (6.4.8), which for the case under discussion takes the form

$$\Delta H^* - T\Delta S^* = - RT \ln \frac{a}{\sqrt{p}}. \qquad (7.3.5)$$

   According to Section 6.4, ΔH^* and ΔS^* are the changes in enthalpy and entropy which occur when 0.5 mole H_2 at 1 atm dissolves, according to equation (7.3.1), in a hypothetical solution of 1 weight$\%$ H in the metal which still obeys Henry's law. Corresponding to this a, the activity of the hydrogen in the metal, may be replaced by its weight percentage as long as the concentration is very small. If the concentrations are so large that appreciable deviations from Henry's law occur, an activity coefficient according to equation (6.4.9) must be introduced. In equation (7.3.5) p is the pressure of the gaseous hydrogen in atmospheres.

   If we represent the concentration (in weight$\%$) of hydrogen in the solution by the symbol c, then equation (7.3.5) for very dilute solutions can be written as

[1]) P. KOFSTAD and W. E. WALLACE, J. Amer. Chem. Soc. **81**, 5019 (1959).
[2]) S. KOMJATHY, J. less-common Metals **2**, 466 (1960).
[3]) P. KOFSTAD, W. E. WALLACE and L. J. HYVÖNEN, J. Amer. Chem. Soc. **81**, 5015 (1959).
[4]) M. W. MALLETT and B. G. KOEHL, J. Electrochem. Soc. **109**, 611 (1962).

$$\log \frac{c}{V_p} = -\frac{\Delta H^*}{4.575\,T} + \frac{\Delta S^*}{4.575} \qquad (7.3.6)$$

or, written differently:

$$\frac{c}{V_p} = \exp\left(\Delta S^*/R\right)\cdot\exp\left(-\Delta H^*/RT\right), \qquad (7.3.7)$$

where, as follows from the above, ΔS^* and ΔH^* are not functions of the concentration or of the gas pressure. It will be seen that these equations not only give the dependence of the solubility on temperature, but also its dependence on pressure. From these equations, one comes to the very important conclusion that the solubility of hydrogen (or any other gas) in a

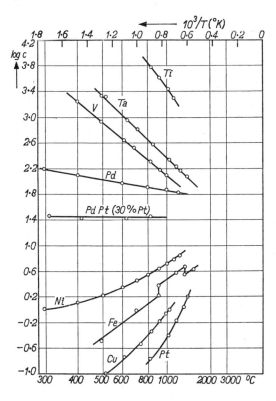

Fig. 31. Logarithm of the solubility of hydrogen in several metals at 1 atm of H_2 as a function of the reciprocal of absolute temperature. The concentration of hydrogen is given in atoms per 10^4 metal atoms (the hydrogen content in cm^3 H_2 per 100 gram metal is obtained from this by multiplication by 2). The data on which the figure is based have mostly been derived from A. Sieverts and co-workers (cf. G. BORELIUS, Ann. Physik **83**, 121 (1927) and Metallwirtschaft, Febr. 1929).

metal at constant pressure will *increase* with the temperature if ΔH^* is positive, i.e. if solution is an endothermic process, while on the other hand, the solubility will *decrease* with rising temperature if ΔH^* is negative, i.e. if solution is an exothermic process. In the case of hydrogen, it is found that the solubility increases with rising temperature in metals such as Cu, Ag, Cr, Mo, W, Fe, Co, Ni, Al, Pt, while the solubility decreases in metals such as Ce, La, Ti, Zr, Hf, Th, V, Nb, Ta, Pd. For a few of these metals Fig. 31 gives the logarithm of the solubility as a function of the reciprocal value of the absolute temperature for a H_2 pressure of 1 atm. The curves have only been drawn for temperatures at which the $\sqrt{p}$ law is approximately valid.

In the simplest case ΔH^* and ΔS^* are virtually independent of temperature. Equation (7.3.6) shows that in that case one should obtain a straight line when $\log c$ is plotted against $1/T$ for a constant value of p. For many dilute solutions of gases in metals this linear relationship is, in fact, found.

Only a small representative selection of the many solutions of gases in metals can be discussed below. In the remaining part of this chapter we shall restrict ourselves to a fairly thorough discussion of the iron-hydrogen and iron-nitrogen systems which are so important in steel technology. A number of other metal-gas systems will be discussed in the following chapter.

7.4. The solubility of hydrogen in iron

When hydrogen reacts with iron no hydrides, but only solutions of the gas in the metal are formed. The solubility of hydrogen in iron, according to the results of many investigators [1-7]), increases with the temperature, which shows (see the preceding section) that the formation of a solution is an endothermic process. Table 12 and Fig. 32 give solubility figures based on the cited literature.

The solubility figures in Table 12 for body-centred cubic α iron have been calculated by the empirical formula of Armbruster (l.c.),

[1]) A. SIEVERTS, G. ZAPF and H. MORITZ, Z. physik. Chem. A**183**, 19 (1938/39).
[2]) E. MARTIN, Arch. Eisenhüttenwes. **3**, 407 (1929/30).
[3]) L. LUCKEMEYER-HASSE and H. SCHENCK, Arch. Eisenhüttenwes. **6**, 209 (1932/33).
[4]) W. GELLER and TAK-HO SUN, Arch. Eisenhüttenwes. **21**, 423 (1950).
[5]) M. WEINSTEIN and J. F. ELLIOTT, Trans. AIME **227**, 382 (1963).
[6]) M. H. ARMBRUSTER, J. Amer. Chem. Soc. **65**, 1043 (1943).
[7]) W. EICHENAUER, H. KÜNZIG and A. PEBLER, Z. Metallkunde **49**, 220 (1958).

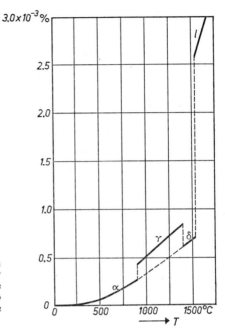

Fig. 32. The solubility of hydrogen in iron at a H_2 pressure of 1 atm in percentages by weight, plotted as a function of temperature in degrees centigrade. α, γ and δ refer to α, γ and δ iron (α and δ bcc, γ fcc). The symbol l indicates liquid iron.

$$\log c\,(H, \alpha) = -\frac{1454}{T} + 1.946 + \tfrac{1}{2} \log p, \qquad (7.4.1)$$

which is based on accurate measurements at 400, 500 and 600 °C. In this formula, c represents the concentration in micromoles hydrogen per 100 gram α iron, T the absolute temperature and p the pressure in mm mercury. This formula also represents the results of other investigators, insofar as they relate to the same or higher temperatures, in a satisfactory manner. This is not true, however, for the values which have been found at temperatures below about 200 °C. But as will be seen later, this (sometimes very large) discrepancy is largely only apparent.

If we express the quantity of hydrogen dissolved in cm^3 H_2 (at 0 °C and 1 atm pressure) per 100 gram iron, then equation (7.4.1) can be written in the following form for a constant H_2 pressure of 1 atm:

$$c\,(H, \alpha) = 54.5 \exp\,(-6650/RT)\ \ cm^3\ H_2/100\ g, \qquad (7.4.2)$$

where R is the gas constant in cal per degree. From equation (7.4.2) we can see that the isothermal solution of hydrogen in α iron according to the reaction

$$\tfrac{1}{2} H_2 \rightarrow [H] \qquad (7.4.3)$$

is accompanied by heat absorption of 6650 cal per gramatom hydrogen (cf. Section 7.3).

Equations (7.4.1) and (7.4.2) are also valid for the solubility of hydrogen in δ iron which is stable between 1400° and 1540 °C and, like α iron, is body-centred cubic.

TABLE 12

SOLUBILITY OF HYDROGEN IN IRON AT AN H_2 PRESSURE OF 1 ATM

Temp. °C	Sol. in cm³ H_2 per 100 gram Fe	Sol. in weight %	Sol. in atom %
20	$6.0 \cdot 10^{-4}$	$5.4 \cdot 10^{-8}$	$3.0 \cdot 10^{-6}$
100	$6.9 \cdot 10^{-3}$	$6.2 \cdot 10^{-7}$	$3.4 \cdot 10^{-5}$
200	$4.6 \cdot 10^{-2}$	$4.1 \cdot 10^{-6}$	$2.3 \cdot 10^{-4}$
300	$1.6 \cdot 10^{-1}$	$1.4 \cdot 10^{-5}$	$7.8 \cdot 10^{-4}$
400	$3.8 \cdot 10^{-1}$	$3.4 \cdot 10^{-5}$	$1.9 \cdot 10^{-3}$
500	$7.2 \cdot 10^{-1}$	$6.5 \cdot 10^{-5}$	$3.6 \cdot 10^{-3}$
700	1.8	$1.6 \cdot 10^{-4}$	$8.9 \cdot 10^{-3}$
900 (α)	3.1	$2.8 \cdot 10^{-4}$	$1.6 \cdot 10^{-2}$
900 (γ)	4.7	$4.2 \cdot 10^{-4}$	$2.3 \cdot 10^{-2}$
1000	5.6	$5.0 \cdot 10^{-4}$	$2.8 \cdot 10^{-2}$
1200	7.5	$6.8 \cdot 10^{-4}$	$3.8 \cdot 10^{-2}$
1400 (γ)	9.3	$8.4 \cdot 10^{-4}$	$4.7 \cdot 10^{-2}$
1400 (δ)	7.4	$6.7 \cdot 10^{-4}$	$3.7 \cdot 10^{-2}$
1540 (δ)	8.6	$7.7 \cdot 10^{-4}$	$4.3 \cdot 10^{-2}$
1540 (liq)	$2.5 \cdot 10^{+1}$	$2.3 \cdot 10^{-3}$	$1.3 \cdot 10^{-1}$
1700	$3.1 \cdot 10^{+1}$	$2.8 \cdot 10^{-3}$	$1.6 \cdot 10^{-1}$

The solubility values in Table 12 for face-centred cubic γ iron, stable between 900° and 1400 °C, have been calculated with the help of the equation of Geller and Tak-Ho Sun (l.c.):

$$\log c\,(\mathrm{H}, \gamma) = -\frac{1182}{T} + 1.677 \quad \text{cm}^3 \ H_2/100 \text{ g.} \qquad (7.4.4)$$

The values for liquid iron are based on experiments of Weinstein and Elliott (l.c.) whose results can be represented by means of the equation

$$\log c\,(\mathrm{H}, \text{liq}) = -\frac{1905}{T} + 2.455 \quad \text{cm}^3 \ H_2/100 \text{ g.} \qquad (7.4.5)$$

If the last two equations are compared with equation (7.3.6), one sees that when hydrogen dissolves isothermally in γ iron and liquid iron quantities of heat, 5400 and 8700 cal/gramatom respectively, are absorbed.

The fact that the solubility of hydrogen in liquid iron is so much greater than in solid iron, as is clearly shown by Fig. 32, has important technical consequences. The sudden drop in solubility on solidification can produce *porosity* in castings and welds. An extreme case is shown in Fig. 33 (see Plate 2 opposite p. 55). This figure shows the pores and cavities in an iron rod which has been obtained by melting pure iron in hydrogen at 1 atm and then pouring it into a water-cooled copper mould [1]).

Of even greater technical importance is the decrease in solubility of hydrogen in solid iron with decreasing temperature. Hydrogen which remains in solution in the metal after solidification can, as a result, be liberated in the form of high-pressure H_2 in all pores and lattice imperfections which have room for hydrogen molecules. This precipitation of H_2 forms the background for all the many unpleasant phenomena which hydrogen produces in steel. We shall discuss the most important of these phenomena in the following, although they will lead us some way outside the domain of thermodynamics. The most serious effect of the precipitation of H_2 at high pressure is the unexpected appearance of ruptures in structures of hydrogen-bearing steel. The excessively high values which are found for the solubility at temperatures below about 200 °C (see above) are due to a large extent to the formation of H_2 in microcavities and to the fact that at low temperatures hydrogen molecules, once liberated, are unable, for kinetic reasons, to return to solution [2]).

7.5. Formation of molecular hydrogen in micro-cavities in iron and steel

In order to clarify what was discussed in the last paragraph of the preceding section, we suppose that a piece of iron at a high temperature, say 1100 °C, is in contact with gaseous hydrogen. Equilibrium will be established in which the quantity of hydrogen dissolved is proportional to the square root of the H_2 pressure. From Table 12 this is seen to be about 6.5 cm^3 H_2 per 100 gram Fe at a hydrogen pressure of 1 atm. Next we suppose that the iron with this hydrogen content is cooled to 20 °C so rapidly that the content immediately after cooling is still unchanged, if not near the surface then at least in the interior of the material. The content of dissolved hydrogen is then about 10^4 times higher than the iron would have in *equilibrium* at 20 °C and 1 atm H_2 (again, see Table 12). Considering the proportionality of the solubility to the square root of the hydrogen pressure this means that the

---

[1]) J. D. FAST, A. I. LUTEIJN and E. OVERBOSCH, Philips Techn. Rev. **15**, 114 (1953).
[2]) The influence of the boundary reactions referred to here and the diffusion of hydrogen in iron are thoroughly discussed in the second part of this work: J. D. FAST, *Interaction of Metals and Gases*, II. *Kinetics and Mechanisms*, to be published.

above-mentioned high content could only be present under equilibrium conditions at an H_2 pressure of 10^8 atm at 20 °C. Since the diffusion coefficient of hydrogen in iron is still fairly large at 20 °C, the dissolved gas will attempt to leave the lattice by diffusion towards the outside surface and to every cavity which is present in the interior of the metal.

This picture still requires a quantitative correction: we have not taken into account the fact that at high pressures hydrogen no longer behaves as a perfect gas. Fig. 34 gives the equilibrium pressures at 20 °C calculated for a number of hydrogen contents in iron and taking into account the deviation from perfect behaviour. We can read off from this figure that in the case discussed the equilibrium pressure is not 10^8, but a few times 10^4 atm. This, however, is still quite a considerable pressure.

In many publications one encounters the supposition that ruptures will occur in hydrogen-bearing iron as soon as the equilibrium pressure under

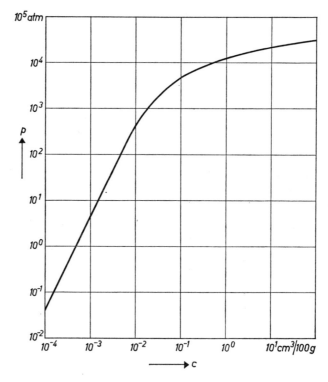

Fig. 34. Equilibrium pressure p of molecular hydrogen as a function of the concentration c of atomic hydrogen in iron, calculated for a temperature of 20 °C taking account of deviations from perfect gas behaviour. (Taken from G. VIBRANS, Arch. Eisenhüttenwes. **32**, 667, 1961).

discussion exceeds the tensile strength of the metal. This supposition must be regarded, in principle. as incorrect, since the hydrogen is practically harmless so long as it remains *in solution* and the equilibrium pressure is only *latent*. In an ideal single crystal of iron even very large hydrogen contents would be unable to produce ruptures. The harmful effects only appear when imperfections are present in the metal (or are introduced by plastic deformation of the metal), which provide an opportunity for the latent pressure to be converted into a real H_2 pressure. Even when imperfections are present fissures do not form immediately, since it takes time to build up the required H_2 pressures.

Hydrogen contents of the order of magnitude of those mentioned above (a few cm^3 H_2 per 100 gram metal) are present in most commercial steels [1]). The example just discussed only differs from reality in that most of the hydrogen in steel does not originate from gaseous hydrogen in the surrounding atmosphere. A much more important source of hydrogen in steel is *water vapour*, which reacts at high temperatures with both solid and liquid iron according to the reaction

$$Fe + H_2O \rightleftharpoons FeO + 2 [H]. \qquad (7.5.1)$$

In the case of the reaction with liquid iron, not only the hydrogen formed, but also the "FeO" formed dissolves wholly or partly in the iron (see Section 4.5). The water vapour which reacts with the iron may come from rust on the scrap used in steelmaking, from components of the slag (lime), from the walls of crucible or furnace and from the gaseous atmosphere. In electric arc welding with coated electrodes, the coating is the chief source of water vapour.

Another, also very important, source of hydrogen in iron and steel is the hydrogen which is formed in *atomic* form at the surface during plating, pickling and electrolysis. The quantity of hydrogen absorbed under these conditions is very much larger than on contact with molecular hydrogen, of which only an extremely small fraction is split into atoms at room temperature.

That high H_2 pressures really do arise in hydrogen-bearing steel, in internal cavities and also at non-metallic inclusions, can be demonstrated by simple experiments [2-4]). Such experiments were carried out as early as 1924. The two following are typical: (1) On the outside of a hollow iron cylinder (with

[1]) See e.g. K. G. SPEITH, H. VOM ENDE and R. SPECHT, Stahl Eisen **82**, 808 (1962).
[2]) C. A. EDWARDS, J. Iron Steel Inst. **110**, 9 (1924).
[3]) P. BARDENHEUER and G. THANHEISER, Mitt. K.W.I. Eisenforschung **10**, 323 (1928).
[4]) C. A. ZAPFFE and C. E. SIMS, Metals and Alloys **11**, 145, 177 (1940) and **12**, 44, 145 (1940).

small internal volume) hydrogen is evolved electrolytically, so that this gas in atomic form penetrates the metal in comparatively large quantities. After some time the pressure inside the bore of the cylinder begins to rise, as can be read from a pressure gauge. In this way one can observe how a pressure of many hundreds of atmospheres is built up, after which the experiment is discontinued for reasons of safety. (2) By means of an acid or by electrolysis atomic hydrogen is evolved inside an open iron pan which is enamelled on the outside. After some time the enamel cracks off the pan — sometimes in a nearly explosive manner — because high-pressure hydrogen has collected at the iron-enamel interface.

Difficulties connected with the last phenomenon sometimes occur in enamelling technique as a result of the reaction, during the enamelling process, between steel and steam according to equation (7.5.1). The steam is liberated from the materials used for enamelling, in particular the "frit" [1]. After cooling, the metal may be strongly supersaturated with hydrogen, which causes serious damage to the enamel layer.

Closely related to the precipitation of hydrogen at high pressure at the boundary between metal and enamel is its formation at non-metallic inclusions in steel. This phenomenon can occur, for example, as a result of the pickling which precedes tin or zinc plating. Part of the atomic hydrogen formed during pickling diffuses inwards and forms H_2 at the inclusions. This may lead to the formation of blisters on the surface, if the inclusions are so close to the surface that the hydrogen pressure is able to push the metal above it outwards by plastic deformation. Fig. 35 shows blisters which have been formed in this way (see Plate 2 opposite p. 55).

Less difficulties would be experienced from hydrogen in steel if one could succeed in producing steel containing a great many tiny pores evenly distributed throughout the metal. In this connection it should be remembered in the first place that the maximum H_2 pressure which can be reached in the pores of the metal is smaller when the relative pore volume is larger and in the second place that the gas pressure is partially compensated by the capillary "counter-pressure" $2\sigma/r$ (cf. Section 6.6). The latter increases as the pores become smaller. It is even better, of course, to prepare the steel by means of vacuum melting or vacuum degassing in such a way that it only contains minimal quantities of hydrogen and, furthermore, to ensure that processes such as pickling which introduce hydrogen at normal temperatures, are avoided.

---

[1] D. G. MOORE, M. A. MASON and W. N. HARRISON, J. Amer. Ceramic Soc. 35, 33 (1952).

7.6. Formation of H$_2$ in iron and steel in imperfections of atomic dimensions

If wires made of pure iron are electrolytically charged with relatively large quantities of hydrogen, then microscopic examination [1,2] shows that after this treatment micro-cracks are present along the grain boundaries (see Fig. 36 on Plate 3 opposite p. 150). The only explanation for their presence is the formation of H$_2$ molecules at these boundaries, giving rise to gas pressures which exceed the local cohesive strength of the metal. The formation of H$_2$ molecules can be expected principally at the large-angle grain boundaries, i.e. at the boundaries between crystals which differ greatly in orientation. Fig. 37a shows a model of one such boundary; Fig. 37b shows a model of a boundary between crystals which differ little in orientation (see Plate 3 opposite p. 150).

The formation of the micro-cracks leads to macroscopically measurable, permanent changes of dimension of the wires. In a particular case, for example, an increase in the diameter of 1.2% was measured, while the increase in length was negligibly small. The relative increase in volume due to the formation of the cracks was in this case thus 2.4%. This formation of cavities produces an increase in the electrical resistance, since a portion of the iron in front of and behind a cavity (seen in the direction of the current) carries no current.

Our experiments showed that after charging, besides the above-mentioned increase in resistance due to part of the iron being excluded from the conduction of current ("shadow effect"), there was also an increase $\Delta\rho$ in the *specific* resistance or resistivity ρ. At first sight it seemed obvious to ascribe this to the presence of dissolved hydrogen in the metal. If, however, the wires were allowed to remain at room temperature until they had lost more than 95% of the absorbed hydrogen, it turned out that $\Delta\rho/\rho$ was still virtually unchanged. Only after the wires had been heated for two hours at 350 °C had $\Delta\rho/\rho$ (measured at 77 °K) decreased from 1.1% to 0.6%. The same effect was found, however, if one started with an entirely hydrogen-free wire which was subjected to plastic deformation such that the same value of 1.1% was obtained for $(\Delta\rho/\rho)_{77}$. In this case, too, heating for two hours at 350 °C produced a reduction of $(\Delta\rho/\rho)_{77}$ from 1.1% to 0.6%.

The increase in the resistivity which is measured after charging with hydrogen must therefore be ascribed to plastic deformation of the metal and not (or only to a very small extent) to dissolved hydrogen. The precipitation of molecular hydrogen under high pressure at the grain boundaries

[1] D. J. VAN OOIJEN and J. D. FAST, Acta Met. **11**, 211 (1963).
[2] J. D. FAST and D. J. VAN OOIJEN, Philips Techn. Rev. **24**, 221 and 252 (1962/63).

thus causes not only micro-cracks, but also a plastic deformation of the metal in the neighbourhood of these fissures.

It is, perhaps, worth while to explain how the effect of the changes in dimension and that of the change in resistivity can be measured separately.

The electrical resistance of a wire without cavities is given by

$$R = \rho \frac{l}{A},\qquad(7.6.1)$$

where ρ is the resistivity, l the length and A the cross-sectional area of the wire. After charging with hydrogen, the resistance can be written:

$$R + \Delta R = (\rho + \Delta\rho)\{l/A + \Delta(l/A)\}.\qquad(7.6.2)$$

The quotient l/A which occurs repeatedly will henceforth be abbreviated to G for "geometry". This will be done, not only to simplify the equations, but also because $\Delta(l/A)$ indicates symbolically the total change in the geometry (external and internal). For the relative increase in resistance we now have in the new notation:

$$\frac{\Delta R}{R} = \frac{\Delta\rho}{\rho} + \frac{\Delta G}{G}.\qquad(7.6.3)$$

It is possible to determine $\Delta\rho/\rho$ and $\Delta G/G$ separately by measuring the resistance before and after charging at two different temperatures T_1 and T_2. Before charging we have

$$R(T_1) - R(T_2) = \{\rho(T_1) - \rho(T_2)\}\, G,\qquad(7.6.4)$$

if G is regarded as being independent of the temperature. This is justified because the change which G undergoes during the transition from T_2 (circa 77 °K) to T_1 (circa 300 °K) as a result of thermal expansion, is very small compared with the change which occurs as a result of hydrogen-charging.

After the wire has been charged with hydrogen we have:

$$R'(T_1) - R'(T_2) = \{(\rho + \Delta\rho)_{T_1} - (\rho + \Delta\rho)_{T_2}\}(G + \Delta G),\quad(7.6.5)$$

where $\Delta\rho$ is the increase in the resistivity, caused by the introduction of hydrogen in the metal and by the plastic deformation as a result of the formation of H_2 at high pressure in the micro-cracks. According to the law of Matthiessen, the increase in the resistivity of a metal, caused by foreign atoms and other lattice imperfections (dislocations, vacancies, etc.) is independent of the temperature. If we assume that this law is also applicable to the problem under consideration (we were, in fact, able to show that this is the case [1]), then (7.6.5) becomes

$$R'(T_1) - R'(T_2) = \{\rho(T_1) - \rho(T_2)\}(G + \Delta G).\qquad(7.6.6)$$

Division of (7.6.6) by (7.6.4) gives:

$$\frac{R'(T_1) - R'(T_2)}{R(T_1) - R(T_2)} = 1 + \frac{\Delta G}{G}.\qquad(7.6.7)$$

With the help of this equation $\Delta G/G$ can be calculated from the experimental results. Since $\Delta R/R$ is also known from the measurements, $\Delta\rho/\rho$ can then be found with the help of equation (7.6.3).

―――――――

[1] D. J. VAN OOIJEN and J. D. FAST, Acta Met. 11, 211 (1963).

We repeat the conclusion which can be reached from the experiments described and also from others not described here [1]): The change in resistance of an iron wire which is measured after charging with hydrogen, is not, or only to a very small extent, caused by dissolved hydrogen. The origin of the change in resistance is rather the permanent damage caused by the hydrogen in the metal in the form of micro-cracks and plastic deformation.

This conclusion is supported by work of other researchers. Of interest, for example, is an X-ray investigation by Tetelman, Wagner and Robertson [2]) into the cause of line broadening, which had been noticed earlier by Bastien and co-workers [3]) after electrolytic hydrogen-charging of iron. Tetelman et al. find that this X-ray line broadening is entirely comparable with that which occurs after cold deformation of a few percent. Also the "recovery" (elimination) of this broadening during heating at 425 °C or 475 °C follows the same pattern in both cases and proceeds with the same activation energy. This energy is much greater than the activation energies of the processes which determine the rate at which hydrogen leaves the metal: the diffusion and surface processes. This shows that the above-mentioned recovery does not depend on the expulsion of hydrogen.

Tetelman and Robertson [4]) showed that hydrogen can also produce micro-cracks and plastic deformation in *single crystals* of iron containing 3% silicon [5]). Their experiments show that this damage occurs after both electrolytic and thermal charging with hydrogen (quenching in water after heating in molecular hydrogen at a temperature of 1000 °C). All the cracks were parallel to cube faces, i.e. to the cleavage planes of the metal. The plastic deformation in the neighbourhood of the cracks was illustrated most beautifully by decoration and etching of the dislocations. There is no doubt that the fissuring and plastic deformation in this case, too, were caused by the precipitation of H_2 at high pressure. It was, however, impossible to discover in which lattice imperfections of the single crystals the precipitation had taken place. It is assumed that this occurred at the incoherent interfaces between the metal and unknown impurities.

It is interesting to note that the occurrence of permanent damage in polycrystalline iron after charging with hydrogen had been deduced as long as

[1]) D. J. VAN OOIJEN and J. D. FAST, Acta Met. **11**, 211 (1963).
[2]) A. S. TETELMAN, C. N. J. WAGNER and W. D. ROBERTSON, Acta Met. **9**, 205 (1961).
[3]) J. PLUSQUELLEC, P. AZOU and P. BASTIEN, Compt. rend. **244**, 1195 (1957).
[4]) A. S. TETELMAN and W. D. ROBERTSON, Trans. AIME **224**, 775 (1962) and Acta Met. **11**, 415 (1963).
[5]) Iron with this silicon content, in contrast to pure iron, has no crystallographic transition points, so that single crystals can be made from it without much difficulty. Furthermore, it is much more brittle than pure iron.

about thirty years ago by Reber [1]) from magnetic measurements. He charged flat rings of soft magnetic iron electrolytically with hydrogen and observed that the magnetic hardness was considerably increased in this way (decrease in the maximum permeability and in the remanent magnetization, increase in the coercivity). The change in magnetic properties remained virtually unaltered after the hydrogen had been expelled.

The cracks which were observed by us to appear along the *grain boundaries* of polycrystalline iron as a result of electrolytic charging with hydrogen, also supply an explanation for several remarkable experiences by other investigators. As an example we mention an extensive study by S. Besnard [2]). She charged iron electrolytically with hydrogen from a bath to which Na₂S had been added as catalyst poison. After charging it was possible to show the presence of *sulphur* along the grain boundaries to deep into the metal. The phenomenon was studied accurately by replacing the sulphur in the Na₂S partially by a radioactive isotope (sulphur 35). However, no satisfactory explanation was given for the penetration of the sulphur. On the basis of our own experiments it would seem obvious to assume that in Besnard's experiments liquid from the electrolytic bath penetrated the metal along cracks at the grain boundaries.

That under extreme conditions hydrogen can cause cracks along grain boundaries in *commercial steel*, was shown long ago by Bardenheuer and co-workers [3]). The authors were able to give an explanation for the damage done to steel upon dip-soldering in molten brass after the steel had been pickled to obtain a clean surface. When the steel is dipped in liquid brass, the hydrogen absorbed during pickling is expelled so rapidly that large cracks form along the grain boundaries. As a result of this, the brass can penetrate deep into the steel (see Fig. 38 on Plate 4 opposite p. 151).

In carbon-bearing iron and steel, under certain conditions, cracks may occur along the grain boundaries due to the formation there of CH_4 at high pressure [4,5]). In technology this phenomenon caused many initial difficulties in carrying out on a large scale a number of syntheses in inorganic and organic chemistry, e.g. in the synthesis of ammonia, methanol and petrol, which requires the application of high pressures

[1]) R. K. REBER, Physics **5**, 297 (1934).
[2]) S. BESNARD, Ann. de Chimie **6**, 245 (1961); S. BESNARD and J. TALBOT, *Colloque sur la diffusion à l'état solide* (organisé à Saclay, 1958), p. 147, North Holland Publ. Co., Amsterdam 1959.
[3]) P. BARDENHEUER and H. PLOUM, Mitt. K.W.I. Eisenforsch. **16**, 129 and 137 (1934); P. BARDENHEUER, Metall **6**, 351 (1952).
[4]) I. CLASS, Stahl Eisen **80**, 1117 (1960).
[5]) L. C. WEINER, Corrosion **17**, 109 (1961).

and temperatures (up to 1000 atm and 600 °C). Nowadays the formation of CH_4 can be successfully avoided by adding alloying elements to the steel which form very stable carbides with carbon.

The formation of CH_4 at high pressure can take place even in the presence of H_2 at *low* pressure (provided that the temperature is not too high). This can be explained by considering the equilibrium

$$Fe_3C + 2 H_2 \rightleftharpoons 3 Fe + CH_4. \qquad (7.6.8)$$

It can be calculated with the help of Table 1 that the equilibrium constant of this reaction,

$$K = p_{CH_4}/p_{H_2}^2,$$

at 300 °C has a value such that an H_2 pressure of 1 atm corresponds to a CH_4 pressure of several thousand atm. At higher temperatures this equilibrium pressure of CH_4 becomes smaller, at lower temperatures larger.

Since the formation of CH_4 described by equation (7.6.8) is an exothermic reaction, it may create the false impression that in a certain temperature range hydrogen dissolves exothermally in steel instead of endothermally.

7.7. Influence of hydrogen on the ductility and on the brittle fracture of iron and steel

It is generally assumed that plastic deformation produces in a metal not only lattice imperfections in the form of dislocations and point defects, but also crack nuclei which, under unfavourable conditions, may extend to real cracks. There are various theories concerning the mechanism of the formation of these nuclei. They have the common feature that each crack nucleus is assumed to originate through the accumulation and coalescence of a number of dislocations under the influence of external shear stresses. In the description which Zener [1]) and Stroh [2]) give of this process, dislocations which move along a slip plane are obstructed by some sort of obstacle (grain boundary or inclusion). The dislocations at the head of such a piled-up group are subjected to great pressure from the following dislocations. They may therefore be forced so close together that they coalesce to a single dislocation with large burgers vector (see Fig. 39). If the burgers vector exceeds a certain size, a wedge-shaped void will have been formed, which can serve as a crack nucleus.

For iron, in which the {110} planes act as slip planes and the {100} planes

[1]) C. ZENER, *Fracturing of Metals*, Am. Soc. Metals, Cleveland (Ohio), 1948, pp. 3-31.
[2]) A. N. STROH, Advanc. Phys. 6, 418 (1957).

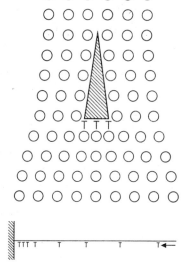

Fig. 39. Illustrating the mechanism by which a crack nucleus forms according to Zener and Stroh. The dislocations move in the direction of the arrow along a slip plane (below) and the first is stopped by an obstacle, such as a grain boundary or inclusion. Under pressure from the following dislocations, a number of them coalesce to form a wedge-shaped void, the crack nucleus (above).

as cleavage planes, Cottrell [1]) has suggested a different mechanism for the formation of crack nuclei. He supposes that dislocations which approach each other along two intersecting slip planes, coalesce along the line of intersection according to the dislocation reaction

$$\frac{a}{2}[\bar{1}\bar{1}1] + \frac{a}{2}[111] \rightarrow a\,[001].$$

In this way a wedge-shaped opening forms between two {100} planes, i.e. in a cleavage direction, and this wedge-shaped opening will be larger as more dislocations are involved in the coalescence (Fig. 40). For a refinement of this theory, the reader is referred to an article by Sleeswijk [2]).

If the iron or steel is supersaturated with dissolved hydrogen then molecular hydrogen at high pressure can form in each crack nucleus. This internal pressure increases the chance of a catastrophic growth of the nucleus into a macro-crack. The adsorption of hydrogen also reduces the surface tension of the iron somewhat, so that the formation of new surfaces (the enlargement of the crack) requires less energy. Besides these two manners in which hydro-

[1]) A. H. Cottrell, Trans. AIME 212, 192 (1958); see also his article in the congress book *Fracture*, Proc. internat. conf. on the atomic mechanisms of fracture, held in Swampscott, Mass., 1959 (editors: B. L. Averbach, D. K. Felbeck and G. T. Hahn).
[2]) A. W. Sleeswijk, Acta Met. 10, 803 (1962).

gen can promote embrittlement, there is, in principle, a third. To appreciate this, one should realize that the formation of a crack nucleus is accompanied by the generation of large shear stresses in the surroundings. These stresses may become so large during the growth of a nucleus that the dislocation sources in the neighbourhood may come into action. If it is fairly easy to activate these sources, the plastic deformation will progress and the stresses will decrease, before macro-cracks form. If, however, the sources are strongly pinned by foreign atoms, there is a probability that cracking will occur before the sources become active. It has been assumed by a number of researchers that the embrittling effect of hydrogen in steel is caused in this way by the interaction of hydrogen atoms and dislocations.

Of the three effects mentioned, however, the formation of molecular hydrogen at high pressure in crack-nuclei is by far the most important [1]). For attempts at a more quantitative treatment of the influence of hydrogen on fracturing of iron and steel, the reader is referred to the literature [2-6]).

If steel contains no hydrogen, the chance of brittle fracturing is greater the higher the rate of deformation and the lower the temperature. The explanation for this lies in the fact that as a rule the dislocations in steel are pinned by foreign atoms (e.g. nitrogen atoms) and that tearing the dislocations loose from these atoms is a thermally activated process.

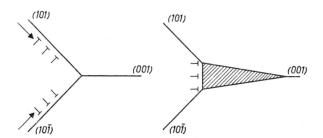

Fig. 40. According to Cottrell crack nuclei form in iron on {100} planes (the cleavage planes) by the coalescence of dislocations moving towards each other along two inter-secting {110} planes (the slip planes).

[1]) J. D. FAST and D. J. VAN OOIJEN, Philips Tech. Rev. 24, 221 and 252 (1962/63).
[2]) F. DE KAZINCZY, J. Iron Steel Inst. 177, 85 (1954).
[3]) N. J. PETCH, Phil. Mag. 1, 331 (1956).
[4]) F. GAROFALO, Y. T. CHOU and V. AMBEGAOKAR, Acta Met. 8, 504 (1960).
[5]) B. A. BILBY and J. HEWITT, Acta Met. 10, 587 (1962).
[6]) K. V. POPOV and V. A. YAGUNOVA, Phys. Met. Metallogr. (transl. from Russ.) 8, 25, no. 2 (1959).

If, however, the steel contains hydrogen, a brittleness may be manifested which is more pronounced as (within certain limits) the rate of deformation is *smaller* and the temperature *higher*. This is now understandable: if hydrogen is to exercise its harmful action by forming high-pressure H_2 in the crack nuclei (produced *during* deformation), the rate of deformation must be low enough and the temperature high enough to give the hydrogen atoms an opportunity to reach the nuclei by diffusion. These hydrogen atoms are perhaps only formed during plastic deformation, from hydrogen molecules which had previously precipitated in pores or other defects. This view is suggested by experiments by Hofmann and collaborators [1,2] concerning the ductility of unalloyed steel (0.22% C) in air and in hydrogen. Tensile test bars of this material, which exhibited great ductility in the normal tensile tests in air, showed only a very small ductility when the experiments were carried out in pure hydrogen at high pressure. The experiments show that the metal absorbs hydrogen during plastic deformation in that gas. On this basis it is an obvious assumption that also high-pressure hydrogen in *pores* can go into solution during plastic deformation, after which it can precipitate in crack-nuclei.

In the following we shall discuss experiments by various investigators which illustrate what has been discussed here. There can be no question, however, of giving even an approximately complete survey of the numerous investigations into the brittle fracturing of steel. The literature on this subject has grown in the last twenty years to unwieldy proportions [3], chiefly as a result of the occurrence of fractures in many welded-steel ships during and after the second world war. In many cases these fractures were of an extremely serious nature, so serious in fact that several ships broke in two as a result. There was, for example, the spectacular case of the tanker Schenectady which in 1943, while lying peacefully at the quay at Portland (Oregon), suddenly broke in two with a bang which was heard several miles away. It is not known whether hydrogen played any part in this case, but it is certainly not inconceivable. The experiments to be discussed will show that the presence of hydrogen in certain types of steel may lead to fracturing after a long delay period if the metal is permanently subjected to external or internal stresses ("delayed fracture").

---

[1] W. HOFMANN and W. RAULS, Arch. Eisenhüttenwes. **32**, 169 (1961).
[2] W. HOFMANN, W. RAULS and J. VOGT, Acta Met. **10**, 688 (1962).
[3] Cf.: M. SZCZEPANSKI, *The brittleness of steel*, John Wiley & Sons, New York, 1963.

LOSS OF DUCTILITY MANIFESTED DURING TESTING

Experiments on the influence of rate of deformation and of temperature on the ductility of hydrogen-bearing steel have been carried out, among others, by Bastien and Azou [1]) and by Baldwin and his co-workers [2]). We shall discuss some experiments by the latter group.

Test bars of a particular type of commercial steel ("spheroidized SAE 1020 steel") were electrolytically charged with hydrogen for one hour in 4% H_2SO_4 (to which a catalyst poison had been added) at a current density of 0.15 A/cm². The hydrogen content after this treatment was about 10 cm³ per 100 gram

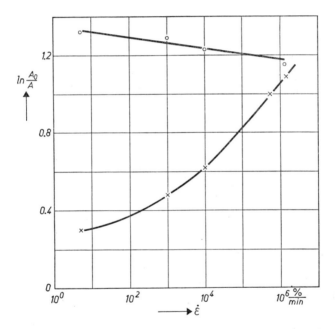

Fig. 41. Influence at room temperature of the strain rate $\dot{\varepsilon}$ on the ductility of uncharged steel (circles) and of hydrogen-charged steel (crosses). As the strain rate increases, the ductility of the uncharged steel decreases, but that of the charged steel rises rapidly. The latter can be explained from the fact that at higher strain rates the hydrogen has less time to diffuse to the crack nuclei.
As a measure of the ductility the true tensile strain ln A_0/A has been plotted as ordinate, A_0 being the initial and A the final cross-section of the test bar at the position of fracture. (After BROWN and BALDWIN, Trans. AIME **200**, 298, 1954.)

[1]) P. BASTIEN and P. AZOU, *Proc. First World Metallurgical Congress* (ed. W. M. Baldwin), p. 535, Am. Soc. Metals, Cleveland (Ohio), 1951; P. BASTIEN, *Phys. Met. of Stress Corrosion Fracture* (ed. T. N. Rhodin), p. 311, Interscience Publ., New York, 1959.
[2]) J. T. BROWN and W. M. BALDWIN, Trans. AIME **200**, 298 (1954); T. TOH and W. M. BALDWIN, *Stress Corrosion Cracking and Embrittlement* (ed. W. D. Robertson), p. 176, John Wiley & Sons, New York, 1956.

metal (about 0.05 atom %). This quantity was not, however, homogeneously distributed over the whole cross-section of the bar. Fig. 41 gives the ductility at room temperature as a function of the strain rate for charged and uncharged

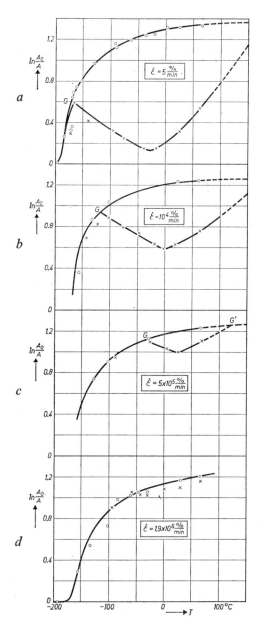

Fig. 42. True tensile strain $\ln A_0/A$ versus temperature at strain rates of 5 %/min (a), 10^4 %/min (b), 5×10^5 %/min (c) and 1.9×10^6 %/min (d). G and G' give the temperatures below which and above which there is no difference in ductility between uncharged steel (circles) and charged steel (crosses). At higher strain rates the region of brittleness, due to the action of hydrogen, becomes smaller. (After BROWN and BALDWIN, Trans. AIME **200**, 298, 1954).

steel. As the measure of ductility, the true fracture strain has been chosen, which is given by the natural logarithm of A_0/A, where A_0 is the original and A the final cross-sectional area of the bar at the point of fracture. At low strain rates the absorbed hydrogen has a marked embrittling effect. If the strain rate is sufficiently large, the charged steel exhibits the same ductility as the uncharged steel.

Fig. 42 gives the ductility of charged and uncharged steel as a function of temperature for four different strain rates. Fig. 43 shows both the influence of temperature and that of the strain rate on ductility, in a for uncharged and in b for charged steel. It will be seen in Fig. 43b that there is a region, viz. surface c, in which the ductility of the charged steel — in contrast to that of the uncharged steel — increases with increasing strain rate and decreases with rising temperature. With the above discussions in mind, this behaviour is understandable. To the left of curve i the temperatures, and with them the diffusion rates of hydrogen, are so small with respect to the strain rates that no appreciable H_2 pressures can form in the crack nuclei during the deformation. To the right of curve i, in surface c, the influence of the hydrogen is more strongly felt as the temperature is higher and the strain rate smaller.

According to Figs 42 and 43 the embrittling action of hydrogen is gradually lost as the temperature rises still further. In surface d (Fig. 43b) the ductility

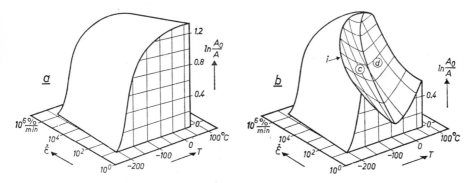

Fig. 43. True tensile strain $\ln A_0/A$ versus temperature T and strain rate $\dot{\varepsilon}$ for uncharged steel (a) and hydrogen-charged steel (b). At the left of curve i the charged steel behaves like uncharged steel, which is understandable since in this region the diffusion rates of hydrogen are low compared with the strain rates and therefore high H_2 pressures have no time to form in the crack nuclei during deformation. At the right of curve i, in surface c, high pressures *are* able to form and brittleness increases with rising temperature. In surface d the brittleness decreases again with rising temperature. This is due to the fact that the equilibrium pressure corresponding to a given hydrogen content decreases rapidly as the temperature increases. Moreover, at these relatively high temperatures the metal rapidly loses its hydrogen. (After TAIJI TOH and BALDWIN, *Stress corrosion cracking and Embrittlement*, Wiley, New York, 1956, p. 176).

not only increases with rising strain rate, but (in contrast to surface *c*) also with rising temperature. This is precisely as expected. In the first place, the equilibrium H_2 pressure corresponding to a particular hydrogen content (the maximum pressure which can be achieved in the crack nuclei) decreases strongly with rising temperature. In the second place, at temperatures above 100 °C the metal loses its hydrogen rapidly.

The effects described have the result that for a given strain rate an appreciable embrittlement will only occur in a limited temperature range. In Fig. 42 the limiting temperatures of the zone of embrittlement correspond to the points *G* and *G'* (the latter point, in Fig. 42*c*, has been obtained by extrapolation). Fig. 44 gives these limiting temperatures as a function of the strain rate for two different hydrogen contents (logarithm of strain rate against reciprocal of the absolute temperature). Only in the area situated between the two branches of the curve can the presence of the hydrogen be noticed.

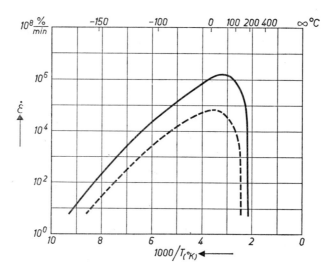

Fig. 44. The limiting temperature below or above which uncharged and hydrogen-charged steel exhibit the same true tensile strain, as a function of strain rate $\dot{\varepsilon}$. For the solid line the charging time at a current density of 0.15 A/cm² was 1 hour, for the broken line 6 minutes. The hydrogen makes its influence felt only in the areas enclosed by the curves. (After Taiji Toh and Baldwin, *Stress corrosion cracking and Embrittlement*, Wiley, New York 1956, p. 176).

DELAYED BRITTLE FRACTURE

Particularly notorious are the fractures that may occur in certain types of chromium-bearing steel with great tensile strength, such as are used in the construction of aeroplanes (e.g. steel H 11 containing 5% Cr and AISI 4340 containing 0.8% Cr). To protect these metals against corrosion they were originally electrolytically coated with a layer of cadmium. The hydrogen which was absorbed by the metal during this plating process, was often the cause of fracture if the metal was subjected to external or internal stresses. It is characteristic of this type of fracture in the first place that it may occur under the influence of a static stress which lies far below the yield point of the metal and in the second place that the fracturing is often preceded by a long delay period. The fractures can be avoided by ensuring that the metal does not absorb any hydrogen, e.g. by applying the cadmium layer by vacuum deposition.

It would seem obvious to suppose that under the influence of the stress small dislocation movements take place, producing crack nuclei in the sense

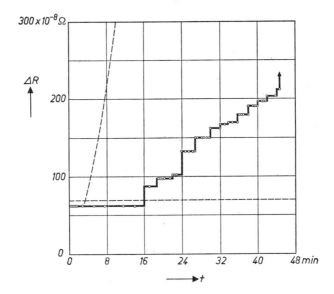

Fig. 45. Resistance increase ΔR of notched steel bars (AISI 4340) after electrolytic charging with hydrogen, as a function of time t at constant load (127 kg/mm^2) and a temperature of -18 °C. After an incubation period of 16 minutes the resistance suddenly increases, presumably owing to the generation of a small crack, after which the resistance stays constant for a while. The phenomenon repeats itself several times before full fracture of the bar (arrow). The broken curves represent parts of the curves from Figs. 46 and 47. (The three figures 45, 46 and 47 are due to STEIGERWALD, SCHALLER and TROIANO, Trans. AIME **215**, 1048, 1959).

of the above. If hydrogen is present this can then enter these nuclei as molecular hydrogen at high pressure, causing them to develop into real cracks.

The phenomena in question have been extensively studied by Troiano and his co-workers. Particularly interesting are experiments in which they measured the electrical resistance of notched test bars (AISI 4340) after these had been deliberately electrolytically charged with hydrogen [1]). During the measurements the bars were subjected to a constant load. Fig. 45 gives the results of the measurements (increase in the resistance as a function of time) for a temperature of $-18\ ^{\circ}$C and a tensile stress of 127 kg/mm² (180,000 psi). When the stress is applied the resistance increases by a certain amount and subsequently remains constant for some time. After this incubation period a sudden increase occurs, followed by a short period of constant resistance. This phenomenon is repeated many times. It may be assumed that each sudden increase in resistance is caused by the growth of a crack nucleus into a real crack of small dimensions.

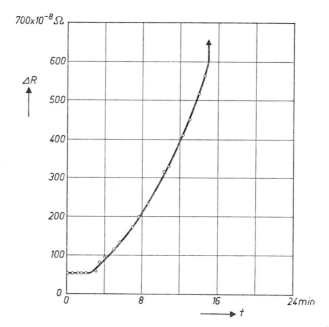

Fig. 46. The same experiment as in Fig. 45, at the *higher* temperature of 27 °C. The incubation period is here much shorter (about 3 minutes) after which there is continuous crack growth.

[1]) E. A. STEIGERWALD, F. W. SCHALLER and A. R. TROIANO, Trans. AIME **215**, 1048 (1959).

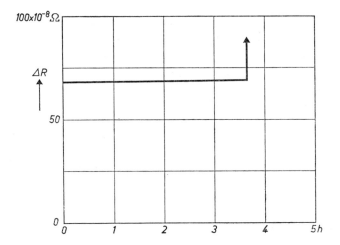

Fig. 47. The same experiment as in Fig. 45, at the *lower* temperature of −46 °C. The incubation period is now much longer, about 3.5 hours, and is followed by a sudden total failure of the bar.

At higher temperature (27 °C) it is noticed that a shorter incubation period is followed by a period of continuous crack growth (Fig. 46), at lower temperature (− 46 °C) on the other hand a longer incubation time is followed by a sudden complete fracture of the bar (Fig. 47). Apparently in the former case the discontinuities have become immeasurably small, while in the latter case the first crack that forms propagates spontaneously through the bar with great velocity.

FLAKES AND WELD CRACKING

A great deal of damage has also been done by hydrogen in the production of large forgings of certain types of low-alloy steel. The damage in question appears on a section through the forging as a multitude of fine cracks, which are described as flakes or hairline cracks. They arise during cooling at temperatures below about 200 °C by the combined influence of thermal stresses or transformation stresses and hydrogen. The greater the hardenability of the steel, the greater the tendency to flaking.

For steels which are susceptible to flaking, the methods of vacuum degassing described in Section 6.6 are of exceptional importance.

In electric arc welding of steel with coated electrodes cracks may occur which are comparable to flakes. They can be prevented by making use of electrodes which only produce very little water vapour during welding.

7.8. Solutions of nitrogen in iron in equilibrium with gaseous nitrogen

In contrast to hydrogen (see preceding sections), nitrogen under suitably chosen conditions will form *chemical compounds* with iron. Of these compounds, two nitrides (Fe_4N and Fe_8N) are of technical importance, particularly in relation to ageing phenomena in steel. Their solubility in steel will be discussed in Section 7.10. This section deals with the solubility of nitrogen in steel in equilibrium with gaseous nitrogen at 1 atm. With the help of the $\sqrt{p}$ law (see Section 7.2) the solubilities at other nitrogen pressures can be calculated approximately from that at 1 atm, provided that the pressures are not so large that serious deviations occur from the behaviour of a perfect gas.

The most accurate method of determining the solubility of nitrogen in iron in equilibrium with an iron nitride or with N_2 is that employing internal friction. It is based on the effect, discovered by Snoek [1]), that dissolved nitrogen at suitably chosen frequencies and temperatures causes internal friction in body-centred cubic iron, which is proportional to the quantity in solution. The same is true, in fact, for all elements which are interstitially dissolved in small concentrations in bcc metals, thus, for example, also for solutions of carbon in bcc iron and for not-too-concentrated solutions of carbon, nitrogen and oxygen in vanadium, niobium and tantalum. We shall give below only the *results* of the measurements of the internal friction and then only where these relate to solutions of nitrogen in iron. For a detailed discussion and explanation of the Snoek effect the reader is referred to the second part of this book [2]).

A complication in the investigation of the iron-nitrogen system is the slowness with which some chemical equilibria are established. If, for example, a pure iron wire is heated in pure nitrogen at a temperature below 900 °C, it turns out that during a period of several hours the metal does not absorb any appreciable quantity of nitrogen. The slowness is not the result of slow diffusion of nitrogen into the metal, but of the sluggishness of the surface reaction

$$N_2 \rightarrow 2\,N_{ads}, \qquad (7.8.1)$$

which of necessity must precede the diffusion into the interior, i.e. the act of solution.

If a small percentage of hydrogen is added to the nitrogen, it is then seen that the equilibrium between metal and gas is much more rapidly established.

[1]) J. L. SNOEK, Physica 8, 711 (1941) and 9, 862 (1942).
[2]) J. D. FAST, *Interaction of Metals and Gases*, II. *Kinetics and Mechanisms*, to be published.

Employing this experimental fact, we have studied the equilibrium between pure iron [1]) and nitrogen at 1 atm in the temperature range 700°-900 °C by adding 1 vol % hydrogen to the nitrogen [2]). The formation of adsorbed nitrogen atoms can then take place via surface reactions which require a smaller activation energy than (7.8.1). In the stated temperature range these may be the reactions

$$N_2 + H_2 \rightarrow 2\,NH_{ads}, \tag{7.8.2}$$

$$NH_{ads} \rightarrow N_{ads} + H_{ads}. \tag{7.8.3}$$

Even in the presence of hydrogen the surface reactions are still relatively so slow that they determine the rate at which nitrogen is taken up by the metal.

The values found for the solubility of nitrogen in α iron are given in Fig. 48. Virtually the same values were found after protracted heating in

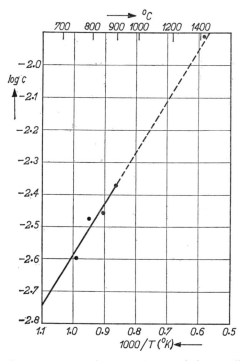

Fig. 48. Equilibrium between gaseous nitrogen at 1 atm and nitrogen dissolved in bcc iron. The figure gives the logarithm of the nitrogen content of the iron in wt% versus the reciprocal of absolute temperature (Fast and Verrijp).

[1]) The pure iron was made by a method described by J. D. FAST, A. I. LUTEYN and E. OVERBOSCH, Philips Tech. Rev. 15, 114 (1953) and J. D. FAST, Stahl Eisen 73, 1484 (1953).
[2]) J. D. FAST and M. B. VERRIJP, J. Iron Steel Inst. 180, 337 (1955).

(99% N_2 + 1% H_2), if wires were used which initially contained *more* nitrogen than corresponded to the solubility. The figure also includes the result of a measurement at 1450 °C, at which temperature iron has the same crystal structure as in the other measurements (iron is body-centred cubic both below 910 °C: α iron, and between 1400 °C and its melting point: δ iron). The straight line in the figure corresponds to the following equation for the solubility of nitrogen in bcc iron in equilibrium with N_2 at 1 atm:

$$c\,(N_2,\ bcc) = 0.098 \exp\,(-7200/RT) \quad \text{weight \% N,} \qquad (7.8.4)$$

or, otherwise expressed:

$$\log c\,(N_2,\ bcc) = -\frac{1575}{T} - 1.009 \quad \text{weight \% N.} \qquad (7.8.5)$$

Equation (7.8.4) or (7.8.5) gives values for the solubility of nitrogen in δ iron which agree reasonably well with those measured in the classical manner by Sieverts, Zapf and Moritz[1]. In this connection "classical" measurements can be taken to include both determinations of solubility by chemical analysis and by measurement of absorbed or liberated gas. For the determination of solubility of nitrogen in fcc iron and liquid iron one is entirely dependent on this sort of measurement. Fountain and Chipman[2], using classical methods, find values of the solubility of nitrogen in α iron which are in good agreement with those given by equation (7.8.4) or (7.8.5). On the other hand, Corney and Turkdogan[3] find somewhat smaller values by classical methods.

The last-mentioned investigators have also determined the solubility of gaseous nitrogen at 1 atm in *fcc iron* (γ iron, stable between 910° and 1400 °C). Their results can be represented by the equation

$$c\,(N_2,\ fcc) = 0.0111 \exp\,(2060/RT) \quad \text{weight \% N} \qquad (7.8.6)$$

or

$$\log c\,(N_2,\ fcc) = \frac{450}{T} - 1.955 \quad \text{weight \% N.} \qquad (7.8.7)$$

These results give satisfactory agreement with those of Sieverts, Zapf and Moritz (l.c.) and with those of Darken, Smith and Filer[4].

Equations (7.8.4) and (7.8.6) show that nitrogen dissolves endothermally

[1] A. Sieverts, G. Zapf and H. Moritz, Z. physik. Chem. A**183**, 19 (1938).
[2] R. W. Fountain and J. Chipman, Trans AIME **212**, 737 (1958).
[3] N. S. Corney and E. T. Turkdogan, J. Iron Steel Inst. **180**, 344 (1955).
[4] L. S. Darken, R. P. Smith and E. W. Filer, Trans. AIME **191**, 1174 (1951).

PLATE 3

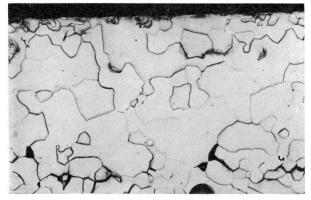

Fig. 36. Soft-annealed pure iron wire after electrolytic charging with hydrogen. Cracks have formed at grain boundaries owing to the evolution of molecular hydrogen at high pressure. The cracks and grain boundaries were made visible by polishing and etching a longitudinal cross-section. Magnification 100×. (Van Ooijen and Fast).

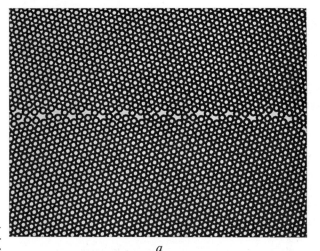

a

b

Fig. 37. (*a*) Model of a high-angle grain boundary, obtained using soap bubbles. The boundary is one between two "crystals" which differ considerably in orientation. The formation of cracks at the boundaries in iron containing hydrogen, shown in Fig. 36, makes it probable that molecular hydrogen forms in the cavities of atomic dimensions existing at the high-angle boundaries.
(*b*) By the soap bubble method it is also possible to bring together two "crystals" with a small difference in orientation. The difference is then bridged by a series of dislocations. (After LOMER and NYE, Proc. Roy. Soc. A **212**, 576, 1952).

PLATE 4

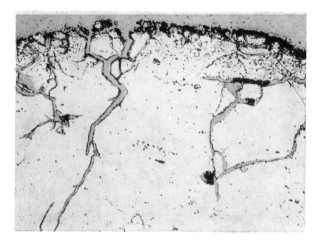

a

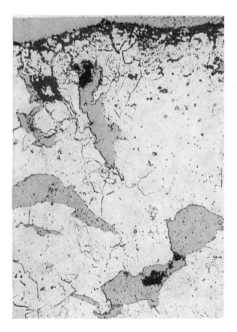

b

Fig. 38. Hydrogen-charged steel wire after dipping in molten brass (magnification 500×). The brass (yellow) penetrates deep into the steel along micro-fissures formed by the rejection of molecular hydrogen at the grain boundaries (*a*). It fills cavities present in the interior of the metal (*b*). (After BARDENHEUER and PLOUM, Mitt. K.W.I. Eisenforsch. **16**, 129 and 137, 1934).

in bcc iron, while in fcc iron it dissolves exothermally. According to the equations the isothermal solution of N_2 as given by the reaction

$$\tfrac{1}{2} N_2 \rightarrow [N] \qquad (7.8.8)$$

is accompanied by the *absorption* of heat of 7200 cal per gram-atom nitrogen in the case of bcc iron, and by the *liberation* of heat of 2060 cal per gram-atom nitrogen in the case of fcc iron (cf. Section 7.3).

The solubility of gaseous nitrogen in *liquid iron* has been determined by many workers. According to an extensive study by Pehlke and Elliott [1] this solubility for a N_2 pressure of 1 atm is given by

$$c\,(N_2, \text{liq}) = 0.0565 \exp(-860/RT) \quad \text{weight\% N} \qquad (7.8.9)$$

or

$$\log c\,(N_2, \text{liq}) = -\frac{188}{T} - 1.248 \quad \text{weight\% N.} \qquad (7.8.10)$$

According to equation (7.8.9) when gaseous nitrogen dissolves isothermally in liquid iron a quantity of heat of 860 cal per gram-atom nitrogen is

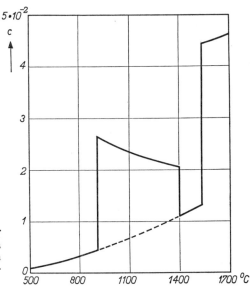

Fig. 49. Solubility (in wt%) of nitrogen at 1 atm in the various modifications of iron as a function of the temperature in degrees centigrade.

[1]   R. D. PEHLKE and J. F. ELLIOTT, Trans. AIME **218**, 1088 (1960).

absorbed. At a temperature of 1600 °C equation (7.8.9) or (7.8.10) gives for the solubility of nitrogen at 1 atm in liquid iron a value of 0.045 weight %, in good agreement with the results of many older investigations [1]. On the other hand, according to the equations mentioned the solubility increases less rapidly with rising temperature than appeared to be the case in the older investigations.

Fig. 49 gives the solubility of nitrogen at 1 atm in the various modifications of iron according to equations (7.8.5), (7.8.7) and (7.8.10). Fig. 50 shows that the solubility of nitrogen in liquid iron obeys the $\sqrt{p}$ law accurately.

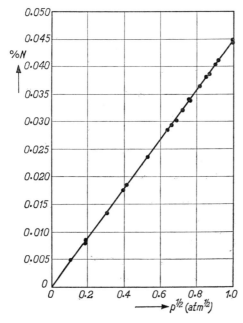

Fig. 50. Solubility of nitrogen in liquid iron at 1600 °C (in wt %) as a function of the square root of the nitrogen pressure .(After Pehlke and Elliott.)

7.9. Reaction of iron with ammonia

If iron is heated at a temperature of, say, 400° or 500 °C in a gas stream consisting of a mixture of ammonia and hydrogen (total pressure 1 atm), then much larger quantities of nitrogen are taken up than when heated in N_2 at the same pressure. Fig. 51 gives, for two different temperatures, the maximum quantity of nitrogen taken up as a function of the NH_3 content

---

[1] See the list of references in the article by R. D. PEHLKE and J. F. ELLIOTT, Trans. AIME **218**, 1088 (1960).

of the gas mixture [1]). At small NH_3 contents, α iron takes up nitrogen in solid solution, the amount dissolved being greater when the gas mixture contains more NH_3. At a particular NH_3 content the solubility limit is exceeded and in addition to the interstitial solution α a new interstitial phase γ' begins to form, with the approximate composition Fe_4N. In compliance with the phase rule, the NH_3 content of the gas remains constant until the α phase has been wholly converted into the γ' phase, which has only a relatively small region of homogeneity. At very high NH_3 contents, eventually, a more nitrogen-rich nitride is formed, known as the ε phase, which cannot be indicated by a simple formula since it has a very extended homogeneity range (see below).

In order to understand the much greater nitriding capacity of NH_3 as compared with that of N_2, we shall consider by way of an example the formation of Fe_4N by the action of NH_3 on α iron. We shall neglect for the present the comparatively small solubility of nitrogen in the metal. In the thermodynamic equilibrium state we should be dealing with the following reactions:

$$8\,Fe + 2\,NH_3 \rightleftharpoons 2\,Fe_4N + 3\,H_2, \qquad (7.9.1)$$

$$2\,NH_3 \rightleftharpoons N_2 + 3\,H_2, \qquad (7.9.2)$$

$$2\,Fe_4N \rightleftharpoons 8\,Fe + N_2. \qquad (7.9.3)$$

The inconvenient decomposition reactions (7.9.2) and (7.9.3), however, proceed so slowly below 600 °C that they do not have to be taken into

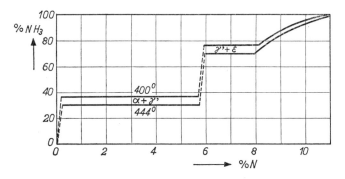

Fig. 51. Reaction of mixtures of hydrogen and ammonia at a total pressure of 1 atm with iron at 400° and 444 °C. The nitrogen content of the iron in wt% has been plotted for the metastable equilibrium state as a function of the composition of the gas in vol %. (According to Brunauer, Jefferson, Emmett and Hendricks).

[1]) S. BRUNAUER, M. E. JEFFERSON, P. H. EMMETT and S. B. HENDRICKS, J. Amer. Chem. Soc. **53**, 1778 (1931).

account when nitriding in a rapid current of gas. On the other hand the metastable equilibrium (7.9.1) is comparatively quickly established. It is therefore possible to determine the equilibrium constant of (7.9.1) for temperatures at which the NH_3 should, in fact, already have dissociated into N_2 and H_2 to within a fraction of a percent at the prevailing pressure or, in other words, at which in the equilibrium state a pressure of thousands of atmospheres nitrogen would have to be applied in order to maintain the NH_3 concentration. This theoretical nitrogen pressure is equal to the dissociation pressure of Fe_4N and can be calculated as follows. The equilibrium constants of (7.9.1), (7.9.2) and (7.9.3) are given by the equations

$$K_1 = \frac{p_{H_2}^3}{p_{NH_3}^2}, \tag{7.9.4}$$

$$K_2 = \frac{p_{N_2} \cdot p_{H_2}^3}{p_{NH_3}^2}, \tag{7.9.5}$$

$$K_3 = p_{N_2}. \tag{7.9.6}$$

From (7.9.4), (7.9.5) and (7.9.6) follows:

$$K_3 = p_{N_2} = \frac{K_2}{K_1}. \tag{7.9.7}$$

At 444 °C equilibrium (7.9.1) corresponding to the first horizontal portion of the nitriding curve in Fig. 51, lies at an NH_3 content of 30% in the gas mixture at 1 atm. At this temperature thus, according to equation (7.9.4) we have: $K_1 = (0.7)^3/(0.3)^2 = 3.81$. The equilibrium constant K_2 is known from the work of Haber and many others and at the given temperature has the value 1.8×10^4. Thus from equation (7.9.7), Fe_4N at 444 °C has a dissociation pressure of about 4700 atm.

> Strictly speaking, this result cannot be wholly correct, since equations (7.9.4) to (7.9.7) are only valid for pressures at which the gases behave as perfect gases. The value calculated is not the pressure but the so-called fugacity of the nitrogen which is in equilibrium with Fe_4N at 444 °C. For the sake of convenience we shall continue to speak of "pressures".

By the method described Emmett et al. [1]) find dissociation pressures for Fe_4N which obey the relationship

---

[1]) P. H. EMMETT, S. B. HENDRICKS and S. BRUNAUER, J. Amer. Chem. Soc. 52, 1456 (1930).

$$p_{N_2}(Fe_4N) = 2.2 \times 10^4 \exp(-2300/RT) \text{ atm.} \qquad (7.9.8)$$

In a similar manner Lehrer [1] finds dissociation pressures for Fe_4N which agree with the values given by equation (7.9.8) in the neighbourhood of 500 °C, but which decrease with decreasing temperature much more rapidly than is the case according to (7.9.8).

The points of intersection of the horizontal portions of any nitriding curve (Fig. 51) with the sloping single-phase lines, indicate the limits of stability of the various phases at the temperature in question. By carrying out the experiments described at a sufficient number of temperatures, one can there- fore find the phase limits as functions of the temperature. This method, combined with X-ray examination, has been used by various research workers to investigate the iron-nitrogen equilibrium diagram.

Fig. 52 gives the diagram for the temperature range 450°-700 °C according to Paranjpe et al. [2]. It differs but little from that compiled at about the same time by Jack [3] and that drawn up twenty years earlier by Lehrer [1]. It will

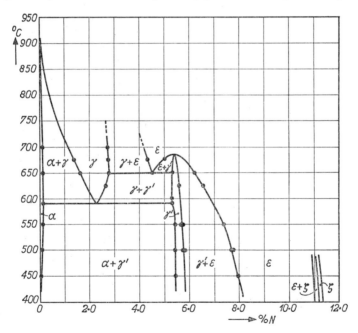

Fig. 52. Equilibrium diagram for the iron-nitrogen system according to Paranjpe, Cohen, Bever and Floe. The horizontal axis gives the nitrogen content in percentages by weight.

[1] E. LEHRER, Z. Elektrochem. **36**, 383 and 460 (1930).
[2] V. G. PARANJPE, M. COHEN, M. B. BEVER and C. F. FLOE, Trans. AIME **188**, 261(1950).
[3] K. H. JACK, Proc. Roy. Soc. A **208**, 200 (1951) and Acta Cryst. **5**, 404 (1952).

be clear from the above that the diagram relates to equilibria which are metastable with respect to nitrogen at 1 atm.

It can be seen from the diagram that the solubility of nitrogen in fcc (γ) iron is much larger than in bcc (α) iron. It is also seen that nitrogen has a stabilizing effect on the γ phase. A content of 2.4 weight % N lowers the transition from fcc to bcc iron from 910° to 590 °C. The nitrogen atoms are distributed in a completely random manner in the octahedral interstices of the fcc lattice.

A face-centred cubic phase again appears at much greater nitrogen contents. This is the γ' phase already mentioned, with a composition of approximately Fe_4N. According to an X-ray investigation by Jack [1]) the nitrogen atoms are here present in a completely ordered manner, such that the octahedral interstice at the centre of each unit fcc cell contains a nitrogen atom, while the octahedral interstices corresponding to the centres of the edges are unoccupied.

In the ε phase, also mentioned earlier, the iron atoms form a hexagonal close-packed structure. The nitrogen atoms are divided amongst the lattice interstices in a disordered manner. The region of homogeneity of the ε phase is very large. At 450 °C it extends from approximately the composition Fe_3N (7.72% N) almost to the composition Fe_2N (11.14% N). In the region of 700 °C, the ε phase can even contain more iron than corresponds to the composition Fe_4N (Fig. 52).

At the extreme right-hand side of the diagram there is an even more nitrogen-rich nitride, the orthorhombic ζ phase, which has a narrow region of homogeneity around the composition Fe_2N [2]). In this phase the interstitial nitrogen atoms are present in an ordered pattern. The packing of the iron atoms deviates only slightly from the hexagonal pattern which exists in the ε phase of nearly the same composition.

Besides the interstitial phases, α, γ, γ', ε and ζ, which appear in the equilibrium diagram (Fig. 52), there are two more interstitial iron-nitrogen phases known which do not exist under stable equilibrium conditions. These are the α' and α'' phases (nitrogen martensite and Fe_8N), which will be discussed in the following sections.

---

[1]) K. H. JACK, Proc. Roy. Soc. A **195**, 34 (1949).
[2]) K. H. JACK, Acta Cryst. **5**, 404 (1952).

7.10. The solubility of nitrogen in bcc iron in equilibrium with Fe₄N or Fe₈N

The most accurate method of determining the solubility of nitrogen in bcc iron in equilibrium with Fe_4N (phase boundary lower left in Fig. 52), is after Section 7.8 the method of internal friction. The measurements are carried out with iron wires with nitrogen contents smaller than 0.1 weight%. We have seen that these can be obtained by heating the wires at a temperature of, say, 580 °C in a gas stream consisting of a suitable mixture of NH_3 and H_2. Contents up to 0.026 weight% can, according to equation (7.8.7), be introduced in the metal via the γ phase by heating the wires at 950 °C in nitrogen at 1 atm (to which 1% H_2 has been added). After rapid cooling all the absorbed nitrogen is still present in the metal.

If, for example, one wishes to determine the solubility of Fe_4N in iron at 400 °C, a wire is employed which contains more nitrogen than corresponds to that solubility and it is kept at the given temperature (in a hydrogen-free atmosphere) long enough for the excess nitrogen to be wholly precipitated as Fe_4N. After quenching in water, the quantity of nitrogen that remained in solution can be determined by measurement of the internal friction.

According to equation (7.9.8), Fe_4N at 400 °C has a dissociation pressure which is not far below 4000 atm. It is not necessary, however, to carry out the heating at 400 °C under this high nitrogen pressure since the equilibrium between Fe_4N and α iron is established fairly rapidly, while the equilibrium with the gas phase is established very slowly. The energy barrier which inhibits the establishment of equilibrium between metal and gaseous phase is the activation energy of the reaction by which nitrogen molecules are formed at the surface by recombination of nitrogen atoms.

Fig. 53 gives the values determined in the manner described for the solubility of Fe_4N in α iron for six different temperatures [1]. They satisfy the equation

$$c \, (Fe_4N, \, bcc) = 12.3 \exp \, (-8300/RT) \quad \text{weight\% N} \qquad (7.10.1)$$

or, otherwise written,

$$\log c \, (Fe_4N, \, bcc) = -\frac{1814}{T} + 1.090 \quad \text{weight\% N.} \qquad (7.10.2)$$

The maximum solubility of nitrogen in α iron in equilibrium with Fe_4N occurs at the eutectoid temperature of 590 °C (Fig. 52) and is seen from the equation to be nearly 0.1 weight% N.

[1] J. D. Fast and M. B. Verrijp, J. Iron Steel Inst. **180**, 337 (1955).

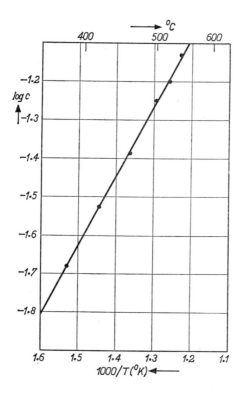

Fig. 53. Solubility of nitrogen in bcc iron in equilibrium with Fe_4N (log [wt % N] versus $1/T$; Fast and Verrijp).

Solubilities which differ only a little from those given by equation (7.10.1) or (7.10.2) have been found by Dijkstra [1]) and by Rawlings and Tambini [2]) using the internal friction method and by Paranjpe et al. [3]) using classical methods.

While making his measurements, Dijkstra discovered a new iron nitride. Using wires containing nitrogen in supersaturated solution, he observed that the precipitation of the excess nitrogen between 200° and 300 °C takes place in two stages (see Fig. 54). The decrease in the quantity in solution quickly comes to an apparent end and microscopic examination shows that it is accompanied by the precipitation of a nitride which, at the time, was unknown. After a longer period at the same temperature there is a further decrease in the amount dissolved which, according to microscopic examination, is accompanied by the disappearance of the first nitride formed and the formation of Fe_4N. Since the new nitride in equilibrium with α iron is

[1]) L. J. DIJKSTRA, Trans. AIME **185**, 252 (1949).
[2]) R. RAWLINGS and D. TAMBINI, J. Iron Steel Inst. **184**, 302 (1956).
[3]) V. G. PARANJPE, M. COHEN, M. B. BEVER and C. F. FLOE, Trans. AIME **188**, 261 (1950).

less stable than Fe_4N, the solubility of nitrogen in α iron in equilibrium with the new phase is greater than its solubility in equilibrium with Fe_4N. These solubilities are given directly by the horizontal portions of curves of the type shown in Fig. 54.

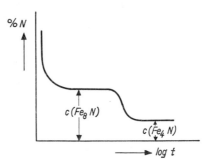

Fig. 54. At relatively low temperatures (e.g. 250° or 200 °C) the precipitation of nitrogen from its supersaturated solution in bcc iron takes place in two steps. The logarithm of the time has been plotted horizontally, the percentage of dissolved nitrogen vertically (Dijkstra).

Research by Jack [1]) has made it seem probable that the new iron nitride is identical with an intermediary phase which occurs in the low-temperature decomposition of nitrogen martensite (see following section). This phase is body-centred tetragonal and can be regarded as a sort of nitrogen martensite with completely ordered nitrogen atoms. It is indicated by the symbol α'' to distinguish it from the ordinary nitrogen martensite which is referred to by the symbol α'. The composition is approximately Fe_8N, but its homogeneity range extends to smaller nitrogen contents. More recent research [2–4]) leaves little doubt that it is indeed identical with Dijkstra's nitride.

The fact that the iron atoms in Fe_8N are arranged in the same pattern as in α iron makes it understandable that the formation of an Fe_8N nucleus in a supersaturated solution of nitrogen in α iron, requires a much smaller activation energy than the formation of an Fe_4N nucleus in which, as we have already seen, the iron atoms have the face-centred arrangement of γ iron. At fairly low temperatures and sufficiently great supersaturation Fe_8N nuclei are therefore formed first. Only much later do the more stable Fe_4N nuclei appear and grow at the expense of the first formed precipitate. Above 300 °C the thermal agitation of the atoms is so intense that the Fe_4N nuclei form without much delay. Below 100 °C the precipitation stops at Fe_8N.

1) K. H. JACK, Proc. Roy. Soc. A **208**, 216 (1951).
2) G. R. BOOKER, J. NORBURY and A. L. SUTTON, J. Iron Steel Inst. **187**, 205 (1957).
3) A. S. KEH and H. A. WRIEDT, Trans. AIME **224**, 560 (1962).
4) K. F. HALE and D. MCLEAN, J. Iron Steel Inst. **201**, 337 (1963).

Fig. 55 shows the results of solubility measurements of nitrogen in α iron in equilibrium with Fe_8N [1]). The solubility as function of the temperature is given according to these measurements by

$$c\,(Fe_8N,\ bcc) = 330\ \exp\,(-9900/RT)\quad \text{weight}\%\ N \qquad (7.10.3)$$

or, otherwise expressed:

$$\log c\,(Fe_8N,\ bcc) = -\frac{2164}{T} + 2.519\quad \text{weight}\%\ N. \qquad (7.10.4)$$

Table 13 gives values of the solubility of nitrogen in α iron in equilibrium with Fe_8N, Fe_4N and N_2 at 1 atm, derived from equations (7.10.4), (7.10.2) and (7.8.5).

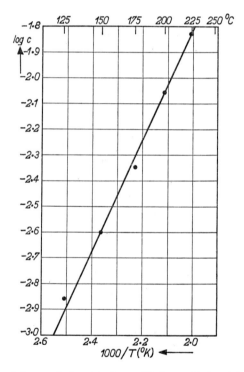

Fig. 55. Solubility of nitrogen in bcc iron in equilibrium with Fe_8N (log [wt% N] versus $1/T$; Fast and Verrijp).

[1]) J. D. FAST and M. B. VERRIJP, J. Iron Steel Inst. **180**, 337 (1955).

TABLE 13

SOLUBILITY OF NITROGEN IN α IRON IN EQUILIBRIUM WITH Fe_8N, Fe_4N and N_2 at 1 atm

Temp. °C	In equilibrium with Fe_8N	In equilibrium with Fe_4N	In equilibrium with N_2, 1 atm
20	$1.4 \cdot 10^{-5}\%$	$7.9 \cdot 10^{-6}\%$	$4.1 \cdot 10^{-7}\%$
100	$5.2 \cdot 10^{-4}$	$1.7 \cdot 10^{-4}$	$5.9 \cdot 10^{-6}$
200	$8.8 \cdot 10^{-3}$	$1.8 \cdot 10^{-3}$	$4.6 \cdot 10^{-5}$
300	$5.5 \cdot 10^{-2}$	$8.4 \cdot 10^{-3}$	$1.7 \cdot 10^{-4}$
400		$2.5 \cdot 10^{-2}$	$4.5 \cdot 10^{-4}$
500		$5.5 \cdot 10^{-2}$	$9.0 \cdot 10^{-4}$
590		$9.7 \cdot 10^{-2}$	$1.5 \cdot 10^{-3}$
700			$2.4 \cdot 10^{-3}$
800			$3.3 \cdot 10^{-3}$
880			$4.2 \cdot 10^{-3}$

7.11. Thermodynamic considerations

In this section we shall discuss some thermodynamic consequences of the previously discussed results obtained from the measurement of the equilibria

$$Fe_8N \rightleftharpoons 8\,Fe + [N], \tag{7.11.1}$$

$$Fe_4N \rightleftharpoons 4\,Fe + [N], \tag{7.11.2}$$

$$\tfrac{1}{2}\,N_2 \rightleftharpoons [N], \tag{7.11.3}$$

where [N] indicates nitrogen dissolved in α iron. From equations (7.10.3), (7.10.1) and (7.8.4) it can be seen that the heats (enthalpies) of these reactions are given by

$$\Delta H(1) = 9900 \text{ cal/g.atom N},$$

$$\Delta H(2) = 8300 \text{ cal/g.atom N},$$

$$\Delta H(3) = 7200 \text{ cal/g.atom N}.$$

Subtracting equation (7.11.3) from equation (7.11.2) gives the dissociation reaction

$$Fe_4N \rightleftharpoons 4\,Fe + \tfrac{1}{2}\,N_2, \tag{7.11.4}$$

with an enthalpy of dissociation

$$\Delta H(4) = 8300 - 7200 = 1100 \text{ cal/mole } Fe_4N.$$

In order to derive the dissociation pressure of Fe_4N as a function of temperature, equation (7.8.4) may be written in a more general form corresponding to equation (7.3.7) and giving the solubility of nitrogen in α iron in equilibrium with N_2 of p atm instead of 1 atm:

$$c\,(N_2,\ bcc) = 0.098\ \sqrt{p_{N_2}}\ \exp\,(-7200/RT)\ \ wt\%\ N. \qquad (7.11.5)$$

If the dissolved nitrogen is in equilibrium not only with the gaseous phase (molecular nitrogen) but also with a Fe_4N precipitate in the metal, then the Fe_4N and the gas phase must also be in equilibrium with each other (Fig. 56).

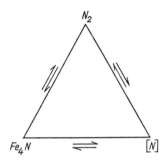

Fig. 56. If a phase A (in our case a solution of nitrogen in iron) is in equilibrium with both a phase B (N_2) and a phase C (Fe_4N), then B and C are also in equilibrium with one another.

From equations (7.11.5) and (7.10.1) therefore follows for the dissociation pressures of Fe_4N (or, strictly, fugacities of N_2 in equilibrium with this compound):

$$0.098\ \sqrt{p_{N_2}}\ \exp\,(-7200/RT) = 12.3\ \exp\,(-8300/RT)$$

or

$$p_{N_2}(Fe_4N) = 1.6 \times 10^4\ \exp\,(-2200/RT)\ \ atm. \qquad (7.11.6)$$

As we have seen in Section 7.9, Emmett et al. derived, from completely different experiments, values for the dissociation pressures of Fe_4N, which agree satisfactorily with this (see equation (7.9.8)).

It is rather surprising that the values of the solubility of Fe_4N in α iron lie quite well on a straight line when plotting $\log c$ versus $1/T$ (Fig. 53). Guillaud and Creveaux [1]) showed that Fe_4N is a ferromagnetic substance with a Curie temperature of 488 °C. This means that below this temperature the internal energy of Fe_4N will show an abnormal decrease with decreasing temperatures. Consequently $\Delta H(2)$ should increase with decreasing temperatures and the $\log c$ versus $1/T$ line should be curved. The straight-line relationship of Fig. 53 could be understood if the magnetic energy of Fe_4N were either relatively small or spread out over a large range of temperatures.

---

[1]) C. GUILLAUD and H. CREVEAUX, Compt. rend. **222**, 1170 (1946).

It could also be understood if the Curie temperature of Fe_4N were either much higher or much lower than 488 °C.

An argument for a Curie temperature lower than about 400 °C could be the large entropy change of the reaction

$$Fe_8N \rightleftharpoons Fe_4N + 4\,Fe \qquad (7.11.7)$$

obtained by subtracting equation (7.11.2) from equation (7.11.1). Equations (7.10.3) and (7.10.1) show that the entropy change is given by

$$\Delta S(7) = R \ln \frac{330}{12.3} \cong R \ln 27.$$

This rather large entropy change could be understood if it were assumed that Fe_8N is below its Curie temperature in the temperature range of solubility measurements (125°-225 °C), whereas Fe_4N is already above its Curie temperature in the range of solubility measurements (380°-580 °C). If this assumption is correct then at low temperatures the enthalpy of solution and the enthalpy of dissociation of Fe_4N will have larger values than those mentioned before (8300 and 1100 cal/mole).

The straight line obtained by plotting log c versus $1/T$ for Fe_8N (Fig. 55) seems to indicate that the composition of this nitride does not change much with temperature when it is in equilibrium with α iron. This need not be in contradiction with Jack's results [1]) showing a large range of compositions as this investigator did not study equilibria.

7.12. Phase transformations in iron-nitrogen alloys

The influence of nitrogen on the properties of iron and steel exhibits great similarity with that of carbon. The similarity consists of a much greater solubility of both elements in fcc iron as compared with that in bcc iron and of the gradual decrease in solubility in bcc iron with decreasing temperature (see with respect to nitrogen Fig. 52). Another similarity is that for kinetic reasons supersaturated solutions of carbon or nitrogen in iron do not precipitate graphite or N_2, as would be the case if equilibrium were fully established, but a carbide or nitride.

If nitrogen dissolved in iron could submit to its thermodynamic tendency to leave the metal in the form of N_2 as the temperature drops (e.g. by

[1]) K. H. JACK, Proc. Roy. Soc. A **208**, 216 (1951).

precipitation in crystal imperfections), then this element would cause similar damage in iron to that done by hydrogen. Only at very high temperature is this the case so that dissolved nitrogen, like hydrogen, can cause porosity during the solidification of liquid iron. In both cases this phenomenon is based on the sudden drop in solubility at the solidification point (see Figs. 32 and 49).

Fig. 57 (see Plate 5 opposite p. 166) shows four lumps of iron obtained by melting (followed by rapid cooling) in a good vacuum, and in argon, hydrogen and nitrogen at 1 atm respectively [1]). As would be expected, in both the vacuum and in argon (which dissolves in neither liquid nor solid iron) a compact lump was obtained, without cavities or pores. In the metal which solidified in hydrogen or nitrogen, on the other hand, large cavities are observed.

In solid iron or steel which after cooling is supersaturated with nitrogen, solid state reactions occur, the nature of which depends on the nitrogen content, the rate of cooling, the initial and final temperatures in the cooling range and the composition of the steel.

THE EUTECTOID REACTION

We shall consider an iron-nitrogen alloy of eutectoid composition (2.4% N). According to the equilibrium diagram in Fig. 52 this alloy above 590 °C consists entirely of homogeneous austenite crystals (γ crystals). If we allow the temperature to drop below 590 °C, the homogeneous γ phase will be transformed (if equilibrium is attained) into a heterogeneous mixture of the α and γ' phases, the first of which (ferrite) contains little nitrogen and the second (Fe_4N) much nitrogen.

The best method to study the progress of the transformation is the isothermal reaction technique, well-known for its results in the study of similar transformations in iron-carbon alloys [2,3]). Working by this method, one determines the transformation rate of austenite at several constant temperatures after rapid cooling from the γ region. The results are registered in a TTT (time-temperature-transformation) diagram.

Fig. 58 gives the TTT diagram determined by Bose and Hawkes [4]) for

[1]) J. D. FAST, Philips Techn. Rev. **11**, 101 (1949).
[2]) E. S. DAVENPORT and E. C. BAIN, Trans. AIME **90**, 117 (1930).
[3]) R. F. MEHL and A. DUBÉ, *Phase Transformations in Solids*, Wiley, New York (1951), p. 545.
[4]) B. N. BOSE and M. F. HAWKES, Trans. AIME **188**, 307 (1950).

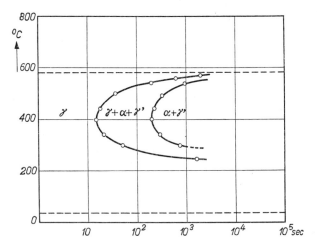

Fig. 58. Isothermal transformation diagram for iron-nitrogen eutectoid alloy (Bose and Hawkes).

the above-mentioned iron-nitrogen alloy of eutectoid composition. The left-hand curve in the diagram gives the periods of time which pass before the transformation begins to a noticeable degree, while the right-hand curve gives the time at which the reaction is practically complete. It will be seen that the reaction only begins after an incubation time which, both close below the eutectoid temperature and below 250 °C, is very long. The greatest transformation rate occurs at about 400 °C where the reaction begins after 15 s and is complete after about 4.5 min. At 150 °C the transformation is not seen to begin even after 12 hours.

The form of the curves in the TTT diagram can be qualitatively understood if one remembers that the reaction in question is a typical nucleation-and-growth process, i.e. a process which requires the formation of *nuclei* of at least one of the new phases and, furthermore, since there is such a big difference of nitrogen content in the newly-formed phases, the *diffusion* of atoms over relatively great distances. Now, nuclei form only slowly at small supercooling, but more rapidly at greater supercooling, while diffusion proceeds rapidly at small and slowly at great supercooling. At both small and large supercooling one can thus expect slow transformation.

The microscopic investigation by Bose and Hawkes (l.c.) shows that the microstructure of the alloys which are transformed below the temperature of the knee of the TTT curve (400 °C) is very different from the micro-structure of the alloys which are transformed above this temperature. This suggests a fundamental difference in the mechanism of transformation above and below the knee as in the case of iron-carbon alloys. It is possible that in the first case the transformation is introduced by the formation of γ' nuclei and in the latter case by the formation of α nuclei in the austenite.

THE AUSTENITE-MARTENSITE REACTION

If an iron-nitrogen alloy of eutectoid composition is cooled rapidly from a temperature in the γ region to 100 °C, then according to Fig. 58 no transformation will occur. The supercooled austenite behaves as though it were a stable phase. However, if this alloy is cooled to below 35 °C, part of the austenite will change into nitrogen-martensite, in which the iron atoms have a similar body-centred arrangement as in ferrite. This transformation, like the analogous one in carbon-bearing iron [1,2]), is brought about by a shear mechanism in which neither long-range nor short-range diffusion plays a role. The martensite formed can therefore be regarded as a strongly supersaturated solution of nitrogen in ferrite, in which the nitrogen atoms are not randomly distributed amongst the interstices, but in such a way that the martensite is body-centred tetragonal instead of body-centred cubic.

The fraction of the austenite which is transformed into martensite is determined solely by the temperature, *not* by the time. If the eutectoid alloy is quenched from 600 °C to 0 °C, for example, then about 10% of the austenite will change into martensite in a fraction of a second, while the rest remains as austenite as long as the temperature is kept at 0 °C. Only when the temperature is lowered further does the quantity of martensite increase. This continues until a temperature of -70 °C is reached. Further lowering of the temperature does not increase the quantity of martensite, although the alloy still contains several tens of percent untransformed austenite. The iron-nitrogen system thus presents an extreme case of the phenomenon of stabilization and retention of austenite.

The temperature M_s at which the transformation of austenite into martensite begins increases with decreasing nitrogen content. It is about 85 °C for 1.85% N.

The lattice parameters of the body-centred tetragonal nitrogen martensite were determined for a number of nitrogen contents by Jack [3]). They are satisfactorily represented by the straight lines in Fig. 59. The corresponding axial ratios c/a, are given as a function of the nitrogen content by Fig. 60.

---

[1]) G. V. KURDJUMOV, J. Iron Steel Inst. **195**, 26 (1960).
[2]) M. COHEN, Trans. AIME **224**, 638 (1962).
[3]) K. H. JACK, Proc. Roy. Soc. A **208**, 200 (1951).

PLATE 5

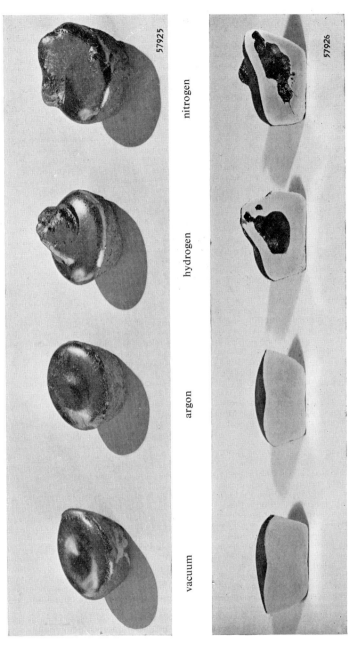

vacuum   argon   hydrogen   nitrogen

a

b

Fig. 57. (*a*) Four lumps of iron, melted and rapidly cooled in a good vacuum, in argon, hydrogen and nitrogen respectively. (*b*) The same lumps, sawn through. The lumps obtained by melting in vacuum and in argon show no cavities or pores, those melted in hydrogen and nitrogen show traces of violent gas evolution during solidification (Fast).

PLATE 6

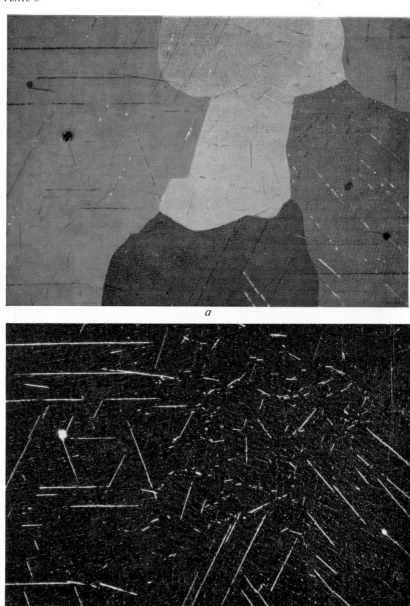

a

b

Fig. 68. Photomicrographs of a precipitate of zirconium hydride in alpha zirconium, made in polarized light (*a*) and with dark field illumination (*b*). Both show the same area of a chemically polished surface (magnification 75×) and are taken from an unpublished investigation by Ir. J. J. de Jong, Philips Research Laboratories, Eindhoven.

The family of {10$\bar{1}$0} planes are the predominant habit planes for the hydride precipitation.

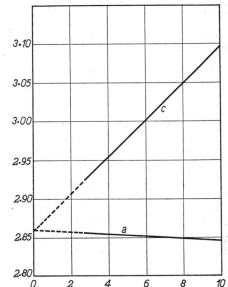

Fig. 59. Unit-cell dimensions of nitrogen-martensites and carbon-martensites according to Jack. Plotted horizontally is the number of interstitial atoms (C or N) per 100 atoms of iron, vertically the unit-cell dimensions in kX units (1 kX unit = 1.002 Ångström).

In both figures the concentrations are given, not in wt% but as the number of interstitial atoms per 100 iron atoms, since the lines then relate not only to the lattice parameters and axial ratios of nitrogen martensites but also, though with rather less accuracy, to those of carbon martensites.

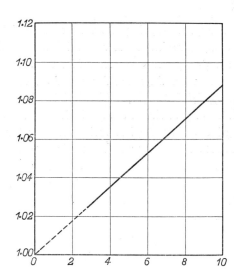

Fig. 60. Axial ratios, c/a, for nitrogen-martensites and carbon-martensites as a function of the number of interstitial atoms (C or N) per 100 iron atoms (Jack).

7.13. Quench ageing

As Table 13 shows, the solubility of nitrogen in α iron in equilibrium with iron nitride at the eutectoid temperature (590 °C) is about 0.1 wt%, while at room temperature it is only about 10^{-5}%. The nitrogen contents of the commercial steels which are used on a large scale for structural purposes lie between these two values: they vary from 0.003 to 0.02%. Since, as we have seen above, equilibrium is not established with the gas phase, the nitrogen present in these steels goes entirely into solid solution at a few hundred degrees centigrade without leaving the metal. On the other hand, in the equilibrium state at room temperature nearly all the nitrogen has precipitated as a nitride.

Since precipitation requires both nucleation and diffusion, relatively rapid cooling from, say, 400 °C to room temperature produces supersaturated solutions of nitrogen in iron. The slow precipitation of Fe_8N which subsequently takes place in the metal, causes a gradual deterioration of the mechanical and magnetic properties, which is known as quench-ageing. The presence of carbon in iron gives rise to similar phenomena since the solubility of this element in α iron, like that of nitrogen, drops sharply with falling temperature.

> Not only after rapid cooling do structural steels exhibit ageing phenomena but also after plastic deformation. This effect, known as strain-ageing, depends on the interaction between the interstitial atoms (C, N) and the dislocations. It will be discussed in the second volume of this work [1].

The quench-ageing of several commercial steels has been studied by many investigators since Köster [2] had first realized that this phenomenon is based on a precipitation process. Fig. 61 shows the results of experiments by Davenport and Bain [3] for a basic open-hearth rimming steel, containing 0.06%C, 0.0042%N, 0.020%O, 0.031%S, 0.012%P, 0.40%Mn and 0.009%Si. The figure shows the gradual, spontaneous change of the Rockwell "B" hardness of the steel at various temperatures after quenching from 720 °C, the eutectoid temperature in the iron-carbon system. It is seen that the metal at 40 °C and higher temperatures during the course of the experiment first becomes gradually harder and then softer. The latter phenomenon is given the name over-ageing. It turns out that the ageing proceeds more swiftly at

[1] J. D. FAST, *Interaction of Metals and Gases*, II. *Kinetics and Mechanisms*, to be published
[2] W. KÖSTER, Arch. Eisenhüttenwes. **2**, 503 (1928-'29), **3**, 553 and 637 (1929-'30)
[3] E. S. DAVENPORT and E. C. BAIN, Trans. Amer. Soc. Met. **23**, 1047 (1935)

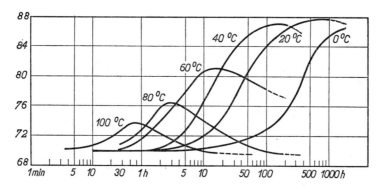

Fig. 61. Quench ageing (at the temperatures given in the figure) of a basic open-hearth rimming steel containing 0.06% C after quenching from 720 °C. On the vertical axis the Rockwell "B" hardness has been plotted, on the horizontal axis the time in minutes up to 60 min. Longer times are given in hours. (Davenport and Bain).

higher temperatures, but the maximum hardness reached is lower at the higher temperatures.

The ageing experiments of other investigators have also been carried out in most cases with commercial steels which, in addition to carbon and nitrogen, always contain oxygen and several other non-metallic impurities. For many years there has been a difference of opinion on the contributions made to the ageing of these steels by carbon, nitrogen and oxygen. This problem could only be solved by studying the process in pure binary alloys, Fe-C, Fe-N and Fe-O. Since the commercial steels in question always contain about 0.5% manganese as the most important metallic impurity, the phenomenon was also studied (after it had turned out that oxygen plays no noticeable part in the ageing) with pure ternary alloys Fe-Mn-C and Fe-Mn-N [1]).

Figures 62, 63 and 64 give the compositions of the alloys studied and the Vickers hardnesses measured. The hardness of each specimen was measured before and after quenching and then after 2 hours heating at 50 °C, 2 hours heating at 100 °C, 2 hours heating at 200 °C and so on. Working in this way, one obtains a general picture of the precipitation phenomena much more quickly than by measuring the hardness as a function of time at a number of temperatures, which is fundamentally the best method.

As would be expected, pure iron, the behaviour of which was determined for comparison, exhibited no quench-ageing (Fig. 62). It is true that the quenching causes internal stresses in the metal which increase the hardness somewhat, but the heating at 50 °C, 100 °C, etc. produce no further change

---

[1]) J. D. FAST, Philips Techn. Rev. **13**, 165 (1951)

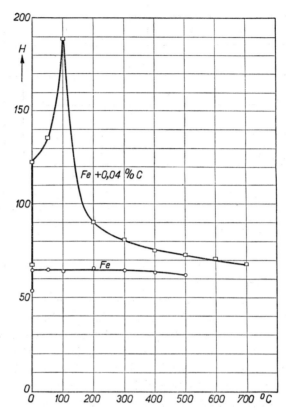

Fig. 62. Vickers hardness of pure iron and iron containing 0.04 wt% carbon before and after quenching in water from 720 °C and after various heat-treatments described in the text (Fast).

in the hardness. It is quite another matter with the iron which contains only carbon or nitrogen. Because the carbon and nitrogen have gone into solution, the hardness after quenching is greater than before. Due to the thermal treatment after quenching, however, the hardness undergoes another important increase. In other words: both the iron containing carbon alone and the iron containing nitrogen alone, exhibit quench-ageing. The maximum hardness is reached after heating at 100 °C. This temperature has no physical significance, considering the arbitrary treatment to which the material was submitted. Iron containing oxygen, according to Fig. 63, only shows a very small quench-ageing and it is not certain whether this can really be attributed to oxygen or not. The small effect due to oxygen, if any, can be explained by the extremely minute solubility of this element in iron, even at such a high temperature as 750 °C.

According to recent measurements by Tankins and Gokcen [1]), the maximum solubility of oxygen in bcc iron is given by the equation

$$\log [\% \text{ O}] = -12630/T + 5.51. \qquad (7.13.1)$$

According to this equation, the solubility of oxygen in delta iron at 1450 °C is about $1.5 \times 10^{-2}\%$ and in alpha iron at 750 °C about $1.5 \times 10^{-7}\%$. Since equation (7.13.1) is the result of measurements on delta iron which is only stable in a temperature range smaller than 140 °C, it can hardly be expected to represent the temperature dependence of the solubility accurately. The extrapolation to the alpha region is therefore rather dangerous.

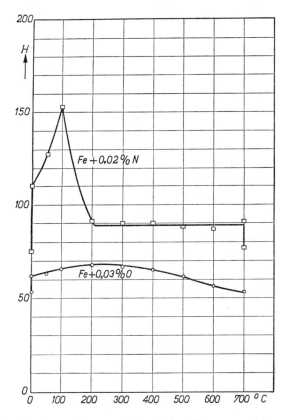

Fig. 63. Vickers hardness of iron containing 0.03 wt% oxygen and iron containing 0.02 wt% nitrogen before and after quenching in water from 750 °C (Fe-O) or 580 °C (Fe-N) and after various heat-treatments described in the text (Fast).

---

[1]) E. S. TANKINS and N. A. GOKCEN, Trans. Amer. Soc. Metals **53**, 843 (1961). In this article many other references will be found to the literature on the solubility of oxygen in iron.

An important result of the experiments is that manganese has little or no influence on the precipitation behaviour of carbon-bearing iron (cf. Figs. 62 and 64), but a great influence on that of nitrogen-bearing iron (cf. Figs 63 and 64). To discover the origin of this different behaviour, the internal friction of the alloys in question was measured as a function of temperature at a frequency of 1 cycle per sec. [1]. From these experiments, which will be discussed in the second volume of this work, it could be concluded that the nitrogen atoms show a preference for the immediate neighbourhood of the manganese atoms. If a nitrogen atom in its diffusion path through the alloy happens to arrive in the neighbourhood of a manganese atom, it remains

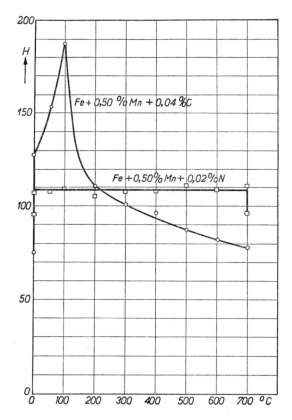

Fig. 64. Vickers hardness of iron with 0.50% Mn + 0.04% C and with 0.50% Mn + 0.02% N before and after quenching in water from 720 °C (Fe-Mn-C) or 580 °C (Fe-Mn-N) and after various heat-treatments described in the text (Fast).

---

[1] J. D. FAST and L. J. DIJKSTRA, Philips Techn. Rev. **13**, 172 (1951)

there for a long time, jumping around this substitutional atom from one energetically favourable interstice to another. This results in enormous delay in the precipitation. It seems obvious to ascribe this behaviour to the fact that nitrogen has a much greater affinity for manganese than for iron. According to Table 1 the standard free enthalpy of formation of Fe_4N is positive ($+2.4$ kcal per mole) while that of $Mn_{2.5}N$ is fairly strongly negative (-23 kcal per mole).

In order to understand the trend in the change of the hardness in Figs. 61, 62, 63 and 64, it is necessary to go into the present views on precipitation in alloys, which, for that matter, has been much more extensively studied with substitutional alloys (especially aluminium containing a few percent copper) than with interstitial alloys. The hardness of a substitutional alloy such as copper-bearing aluminium, after quenching from a temperature at which the alloy is homogeneous, also increases at first at a constant temperature (e.g. 100 °C), after that it decreases. On the basis of the equilibrium diagram one would expect the precipitation of the compound Al_2Cu. Merica, Waltenberg and Scott believed in 1919 that the hardness changes could be explained by assuming that at first a very finely divided precipitate of Al_2Cu formed which later gradually coalesced. The maximum hardness would be reached at a "critical degree of dispersion" of the Al_2Cu, which was below microscopic resolution. Later it turned out that the precipitation in copper-bearing aluminium takes place in a much more complicated manner and that one is not, as in the old view, dealing with a markedly heterogeneous system right from the start of the precipitation, but with a system in which the degree of heterogeneity gradually increases [1,2]. In the "pre-precipitation" stage, domains are formed with a composition which deviates from that in the homogeneous alloy and from that in the stable precipitate. These regions are coherent with the matrix and are the cause of strains in this matrix which explain the increased hardness [3]. For the sake of convenience these regions are usually indicated by the name "precipitate" and one speaks of coherent precipitation. As time passes the composition and the structure of the precipitate change and the coherence with the matrix is lost. The incoherent precipitate causes a much smaller hardness, which is eventually reduced still further by coalescence.

In iron-nitrogen alloys, too, a coherent precipitate forms first, the above mentioned Fe_8N. The reason that this occurs, although the free energy will

[1] A. GUINIER, Physica 15, 148 (1949)
[2] G. D. PRESTON, Phil. Mag. 26, 855 (1938)
[3] A. H. GEISLER, *Phase Transformations in Solids*, Wiley, New York (1951), p. 387

eventually decrease more when Fe_4N is formed, is partly the same as the reason why the stable Al_2Cu is not directly formed in copper-bearing aluminium: due to the coherence with the matrix the interfacial energy is only small and therefore the formation rate of the less stable precipitate is greater than that of the stable, non-coherent precipitate. Furthermore, the formation of Fe_4N, in contrast to that of Fe_8N, involves a regrouping of the iron atoms (see Section 7.10).

The fact that in iron-nitrogen alloys (with the experimental method described) no hardening is observed when the metal contains 0.5% manganese (Fig. 64) can be explained from the above. In order to form a precipitate, the nitrogen atoms must be able to break away from the manganese atoms around which, as we have seen, they jump about. In the short period of the experiments they can only do this at temperatures which are so high that already Fe_4N is mainly formed instead of Fe_8N and that also the coalescence of the precipitate progresses rapidly.

The retarding influence of manganese on the precipitation of nitrogen has the effect that in commercial mild steel (which always contains manganese and nitrogen) quench-ageing phenomena can occur even after slow cooling, since this "slow" cooling still proceeds "quickly" when compared with the rate of precipitation of iron nitride. Among these phenomena there is not only the gradual increase of the mechanical hardness but also that of the magnetic hardness. From the work of various investigators, particularly that of Köster [1]), it was already known for a long time that mild steel (also slowly cooled mild steel) need contain only 0.005% nitrogen for its coercivity to be doubled when heated at 100 °C for several hundred hours. This increase in coercivity is due to the precipitation of the nitrogen in the form of nitride [2]). From the experiments described in the foregoing it may be concluded that the phenomenon of slow magnetic ageing is to be ascribed not to the presence of nitrogen alone but to the simultaneous occurrence of manganese and nitrogen in the steel.

Before silicon steel was used on a large scale as a magnetically soft material, a great deal of inconvenience was caused in electrical engineering by the ageing of steel. Silicon steel, besides its well-known favourable properties, has the additional advantage that it shows virtually no ageing since all the nitrogen which is present in it is chemically bound in the form of Si_3N_4, a nitride which has a great stability (see Table 1).

---

[1]) W. Köster, Z. anorg. allg. Chem. **179**, 297 (1929)
[2]) For the influence of a precipitate of iron nitride on the coercivity see also: J. Kerr and C. Wert, J. appl. Phys. **26**, 1147 (1955) and W. Köster and L. Bangert, Arch. Eisenhüttenwes. **25**, 231 (1954)

CHAPTER 8

SOLUTIONS OF GASES IN METALS - II

8.1. The zirconium-hydrogen and titanium-hydrogen systems

The related elements zirconium and titanium are among the metals which have only achieved a great technological importance in the last decennia, zirconium through the development of nuclear energy and titanium due to the requirements of high speed and space flight. The mechanical properties of both metals are satisfactory only when their oxygen, nitrogen and hydrogen contents are very small. The same is true for zirconium-base and titanium-base alloys.

In this section we shall restrict ourselves to a discussion of the Zr-H and Ti-H systems, which exhibit a great mutual similarity. They form an interesting contrast to the Fe-H system discussed in the previous chapter. In the Fe-H system the affinity between metal and hydrogen is only slight: they form no chemical compounds and the gas dissolves in the metal endothermally and in comparatively small quantities (cf. Fig. 31 and Table 12). In the Zr-H and Ti-H systems, on the other hand, the mutual affinity is great, chemical compounds are formed and the gas dissolves in the metal exothermally and in relatively large quantities.

The maximum quantity of hydrogen which can be absorbed by the two metals corresponds approximately to the composition ZrH_2 and TiH_2 respectively. In practice a maximum hydrogen-to-metal ratio is found which is always smaller than 1.99. The best-studied system is zirconium-hydrogen, the phase diagram for which is given in Fig. 65, which is based on the results of several studies [1-11]. The titanium-hydrogen system is of the same

[1] E. A. GULBRANSEN and K. F. ANDREW, Trans. AIME 203, 136 (1955)
[2] R. K. EDWARDS, P. LEVESQUE and D. CUBICCIOTTI, J. Amer. Chem. Soc. 77, 1307 (1955)
[3] C. E. ELLS and A. D. McQUILLAN, J. Inst. Metals 85, 89 (1956)
[4] D. A. VAUGHAN and J. R. BRIDGE, Trans. AIME 206, 528 (1956)
[5] M. W. MALLETT and W. M. ALBRECHT, J. Electrochem. Soc. 104, 142 (1957)
[6] T. B. DOUGLAS, J. Amer. Chem. Soc. 80, 5040 (1958)
[7] L. D. LA GRANGE, L. J. DIJKSTRA, J. M. DIXON and U. MERTEN, J. phys. Chem. 63, 2035 (1959)
[8] V. V. SOF'INA, Z. M. AZARKH and N. N. ORLOVA, Soviet Physics, Crystallography 3, 544 (1959). Translated from Russian.
[9] L. ESPAGNO, P. AZOU and P. BASTIEN, Compt. rend. 247, 1199 (1958); 248, 2003 (1959) and 249, 1105 (1959)
[10] G. ÖSTBERG, J. Nucl. Mat. 5, 208 (1962)
[11] G. G. LIBOWITZ, J. Nucl. Mat. 5, 228 (1962)

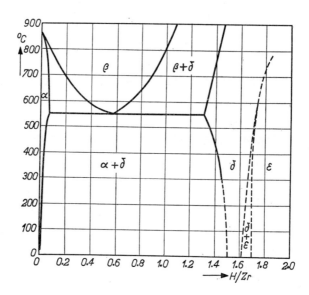

Fig. 65. Equilibrium diagram for the zirconium-hydrogen system. The equilibria do *not* relate to one particular hydrogen pressure. The hydrogen content of the alloys is given as atomic ratio H/Zr.

eutectoid type [1-5]). With rising temperature both metals undergo a phase transformation from the hexagonal close-packed to the body-centred cubic modification (Zr at 865 °C, Ti at 885 °C). The modification which is stable at low temperatures is referred to as the α phase, the other as the β phase.

As can be seen from Fig. 65, the solubility of hydrogen in α Zr (hcp) is much smaller than in β Zr (bcc). At the eutectoid temperature (550 °C) the solubility in α Zr is about 6 atom%, in β Zr about 37 atom%. (In the Ti-H system the eutectoid temperature is about 300 °C, while the solubilities mentioned are about 8 and 40 atom% H). In addition to the solid solutions, α and β, with the crystal structures of pure zirconium, two hydrides with a fluorite type structure occur in the Zr-H system: the cubic δ hydride and the tetragonal ε hydride both of which have an extensive range of homogeneity (Fig. 65) [6]. They can be described as solid solutions of hydrogen in cubic and tetragonal face-centred zirconium, in which the hydrogen atoms are

[1] A. D. McQUILLAN, Proc. Roy. Soc. A204, 309 (1950)
[2] G. A. LENNING, C. M. CRAIGHEAD and R. I. JAFFEE, Trans. AIME 200, 367 (1954)
[3] S. S. SIDHU, L. HEATON and D. D. ZAUBERIS, Acta Cryst. 9, 607 (1956)
[4] L. D. JAFFE, Trans. AIME 206, 861 (1956)
[5] W. KÖSTER, L. BANGERT and M. EVERS, Z. Metallkde 47, 564 (1956)
[6] For the occurrence of a metastable zirconium hydride see: R. L. BECK, Trans. Amer. Soc. Metals 55, 542 (1962)

present in the tetrahedral interstices [1]). A similar description is valid for titanium-hydrogen. On this basis it seems obvious to suppose that, also in the hcp α phases of Zr and Ti, the hydrogen atoms occupy the tetrahedral interstices and not, like oxygen and nitrogen atoms, the much larger octahedral interstices.

From the phase diagram it can be seen that the solubility of hydrogen in both modifications of zirconium (the same is valid for titanium) increases with rising temperature. Dissolution thus takes place endothermally. This is not contrary to the exothermal solution of hydrogen in zirconium, stated in Section 7.3 and at the beginning of this section, since that statement referred to the solution of *gaseous* hydrogen at constant pressure. In the phase diagram, on the other hand, we are dealing with the solubility of hydrogen in zirconium in equilibrium with zirconium hydride or more simply: the solubility of δ hydride in the metal. One can visualize that the solution is formed in two steps: in the first step δ hydride is converted into α phase and hydrogen until 0.5 mole H_2 has been formed, then the gas is dissolved in (an infinite quantity of) α:

$$\delta \rightleftharpoons \alpha + \tfrac{1}{2} H_2, \tag{8.1.1}$$

$$\tfrac{1}{2} H_2 \rightleftharpoons [H]_\alpha, \tag{8.1.2}$$

adding $\qquad\qquad \delta \rightleftharpoons \alpha + [H]_\alpha. \tag{8.1.3}$

In these equations the symbols δ and α refer to the compositions of these phases, as given by the phase boundary curves at the temperature in question.

The fact that reaction (8.1.3) is endothermal implies that when reaction (8.1.1) proceeds isothermally more heat is absorbed from the surroundings, than is given out to the surroundings when reaction (8.1.2) proceeds isothermally. According to Gulbransen and Andrew, l.c., the heat of reaction (8.1.1) is 22875 cal and that of reaction (8.1.2) is — 14250 cal per gramatom of hydrogen. As a consequence the heat of reaction (8.1.3) is given by $22875 - 14250 = 8625$ cal per gramatom of hydrogen. It is the heat of solution of a quantity of δ containing 1 gramatom of hydrogen in (an infinite quantity of) α.

The heats of reactions (8.1.1) and (8.1.2) were experimentally determined by measuring the decomposition pressures of a number of Zr-H alloys as functions of the temperature. If for one of these alloys (containing, say, 3 atom% hydrogen) $\log p(H_2)$ is plotted against $1/T$ two straight lines are obtained like those schematically represented in Fig. 66. Line AB relates to

---

[1]) See also: R. E. RUNDLE, C. G. SHULL and E. O. WOLLAN, Acta Cryst. 5, 22 (1952)

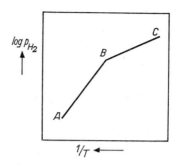

Fig. 66. Schematic logarithmic plot of the decomposition pressure of hydrogen versus $1/T$ for a Zr-H alloy containing less than 6 at % hydrogen. In the temperature range corresponding to the straight line AB the α and δ phases (see Fig. 65) are co-existent, at higher temperatures only the α phase is stable. The slope of AB gives the heat of reaction of 0.5 mole H_2 with a quantity of α such that δ is produced (α and δ with the co-existing compositions). The slope of BC gives the heat of solution of 0.5 mole H_2 in the α phase. Point B indicates the temperature at which zirconium is wholly in the α state and saturated with hydrogen.

the temperature range in which the α and δ phases are co-existent (see Fig. 65), line BC to the temperature range in which only the α phase is stable. The slope of AB gives the heat of reaction (8.1.1), in which α and δ have the co-existent compositions. The slope of BC gives the heat of solution of 0.5 mole H_2 in an infinite amount of α (reaction (8.1.2)). It will be obvious that point B indicates the temperature at which the metal consists wholly of the α phase saturated with hydrogen, i.e. it gives the temperature at which the overall composition concerned is identical to the solubility limit of hydrogen in α Zr.

Gulbransen and Andrew, l.c., also used another method to determine this solubility limit (terminal solubility). Here, the decomposition pressures at a given temperature were plotted as a function of the square of the concentration. In agreement with the $\sqrt{p}$ law (Section 7.2), a straight-line relationship is found. As soon as the solubility limit is reached, the pressure remains constant with increasing concentration (see Fig. 67 and compare this with Fig. 10).

From the results of the authors mentioned, the solubility limit of H in α Zr (i.e. the above-mentioned solubility of hydrogen in α Zr in equilibrium with δ hydride) in the temperature range 350°-550 °C is given to a good approximation by

$$c = 1170 \exp\,(-8625/RT)\quad \text{atom}\,\%\ \text{H} \qquad (8.1.4)$$

or, in a different form:

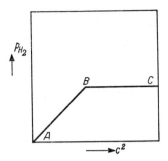

Fig. 67. Equilibrium H_2 pressure of Zr-H alloys at a constant temperature below 550 °C as a function of the square of the hydrogen concentration in the metal. The solubility limit is reached at point B.

$$\log c = -\frac{1885}{T} + 3.068 \quad \text{atom\% H.} \qquad (8.1.5)$$

The solubilities given by these formulae agree satisfactorily with those determined by other investigators, e.g. with those of Mallett and Albrecht (l.c.) and of Espagno, Azou and Bastien (l.c.).

If the concentration of a dissolved substance is very small, it is often expressed in ppm = parts per million by weight = 10^{-4} wt%. If the number of atom% H is indicated by the symbol a and the atomic ratio H/Zr by the symbol x, then the hydrogen content of zirconium is given by:

$$\text{ppm} = \frac{110\,a}{1 - 0.01\,a} = \frac{1.1 \times 10^4\,x}{1 + 0.011\,x}.$$

When equation (8.1.5) is used to calculate the solubility of hydrogen in zirconium at 20 °C, one finds $c = 4.3 \times 10^{-4}$ atom% = 0.05 ppm. Although this extrapolation to room temperature is perhaps not wholly justifiable, one can certainly conclude from it that the solubility of H in α Zr at that temperature is extremely small and certainly lies below 1 ppm. The same is true for the solubility of hydrogen in alpha titanium.

8.2. Embrittlement of zirconium and titanium by hydrogen

Hydrogen is not only one of the most dangerous impurities in iron and steel (see the previous chapter), but also in zirconium and titanium and in zirconium-base and titanium-base alloys. One of the ways in which hydrogen can be absorbed by these metals is by corrosion in high-temperature water and steam. Due to the rapid decrease in solubility of the hydride with decreasing temperature, the hydrogen content is nearly always greater than the metal can hold in saturated solution at ordinary temperatures. Even comparatively small hydrogen contents (a few tens of ppm or more, depending on the composition of the metal) are therefore sufficient to cause embrittlement. If the excess hydrogen before testing is present in the form of hydride, impact embrittlement is observed which becomes more severe with increased strain rate and decreased temperature. When, on the other hand, the excess hydrogen before testing is present in supersaturated solution in the metal, low-strain-rate embrittlement is observed.

Low-strain-rate embrittlement depends, as in the case of iron containing hydrogen (Section 7.7), on the rejection of hydrogen from the supersaturated solution during plastic deformation. The mechanism in the case of zirconium and titanium, however, is completely different to that for iron. While in iron the hydrogen rejection occurs by precipitation of molecular hydrogen, in zirconium and titanium it occurs by precipitation of a crystalline, brittle hydride (the hydride which is referred to in the Zr-H system, Fig. 65, as the δ phase).

There are indications that significant supersaturation with hydrogen never occurs in pure zirconium and pure titanium [1]. Even after very rapid cooling of hydrogen-bearing metal, the hydride is probably already present as fine particles which tend to coalesce with time. When the metal is cooled slowly, the hydride precipitates in the form of platelets which, for the larger part, lie parallel to the crystallographic planes of the $\{10\bar{1}0\}$ type [2-4], i.e. parallel to the principal slip planes in titanium and zirconium. Fig. 68 (opposite p. 167) shows the presence of a hydride precipitate in zirconium [5].

In contrast to what is valid for the pure metals, in many alloys of titanium and zirconium it is possible for relatively large supersaturation with hydrogen to occur, causing low-strain-rate embrittlement. This type of hydrogen embrittlement is most often observed in α/β titanium alloys, which are alloys in which the hcp α phase and the bcc β phase co-exist. An example of this type of alloy is titanium containing 8% vanadium, as can be seen from the transformation point diagram of the Ti-V system (Fig. 69). The solubility

[1] See e.g. D. N. WILLIAMS and R. I. JAFFEE, J. less-common Met. 2, 42 (1960) and D. G. WESTLAKE, Acta Met. 12, 1373 (1964).
[2] J. P. LANGERON and P. LEHR, Revue Métall. 55, 901 (1958), Zr-H.
[3] D. G. WESTLAKE and E. S. FISHER, Trans. AIME 224, 254 (1962), Zr-H.
[4] TIEN-SHIH LIU and M. A. STEINBERG, Trans. Amer. Soc. Metals 50, 455 (1958), Ti-H.
[5] According to J. E. BAILEY, Acta Met. 11, 267 (1963) electron and X-ray diffraction patterns show that the hydride which precipitates in zirconium with low hydrogen concentration is not purely cubic but slightly tetragonal. It can be regarded as a distorted state of the δ fcc phase.

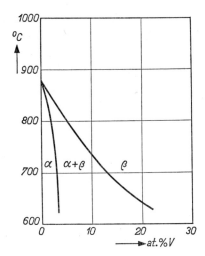

Fig. 69. Part of the titanium-vanadium equilibrium diagram.

of hydrogen in the β phase is much greater than that in the α phase, but it also decreases strongly with decreasing temperature and is probably very small at room temperature.

Fig. 70 gives a schematic picture of the influence of strain rate and temperature on the tensile ductility of a typical α/β titanium alloy containing 375 ppm hydrogen [1]. As the figure shows, embrittlement is absent at high strain rates, so that one is clearly dealing with low-strain-rate embrittlement

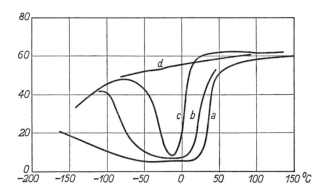

Fig. 70. Influence of strain rate and temperature on the tensile ductility of a typical α/β titanium alloy containing 375 ppm hydrogen. The tensile ductility (reduction in area, per cent) has been plotted as ordinate. The strain rates are: (a) 0.005, (b) 0.05, (c) 0.5 and (d) 1.0 inch/minute (Williams).

[1] D. N. WILLIAMS, J. Inst. Metals 91, 147 (1962-1963)

and *not* with impact embrittlement, which only occurs at much greater hydrogen contents in the alloys in question. Fig. 71 shows, for the same alloy, the influence of hydrogen content on the tensile ductility at the lowest applied strain rate of 0.005 inch/minute [1]).

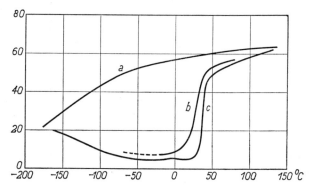

Fig. 71. Influence of hydrogen content and temperature on the tensile ductility of a typical α/β titanium alloy at a strain rate of 0.005 inch/minute. The tensile ductility (reduction in area, per cent) has been plotted as ordinate. The hydrogen contents are: (a) 20, (b) 250 and (c) 375 ppm (Williams).

Precipitation of hydride from the supersaturated solution of hydrogen in the metal can only occur during plastic deformation if this takes place so slowly that there is sufficient time for the formation of hydride nuclei and for the diffusion of hydrogen to these nuclei. The diffusion is unable to support the growth (during the test) of the hydride to particles of dangerous dimensions if the strain rate is relatively large or the temperature relatively low. In agreement with this, Figs. 70 and 71 show that the ductility of the alloy in question at great strain rate is virtually independent of the hydrogen content (to 375 ppm) and at very low temperatures is virtually independent of hydrogen content and strain rate (in the range 0.005 to 1.0 inch/minute).

Even with relatively great supersaturation and diffusion rate of hydrogen in the metal, embrittlement can only occur if the nucleation rate of the hydride is not too small. The nucleation rate is determined not only by the deformation stresses, but also by the temperature and the interface energy of hydride and metal. As the temperature rises, the supersaturation with hydrogen becomes smaller as a result of the increasing solubility, which implies that the decrease in free energy per mole precipitate formed becomes smaller. It is thus more difficult for hydride nuclei to form as the temperature rises. It would therefore seem obvious to suppose that the disappearance of the influence of hydrogen on the ductility with rising temperature (Fig. 70)

---

[1]) D. N. WILLIAMS, J. Inst. Metals **91**, 147 (1962-1963)

results from the retardation or even the complete absence of nucleation. It is also to be expected that nuclei are formed in the α and β phases at very different rates. We have already seen in the preceding section that the metal atoms in the hydride are arranged in a face-centred cubic lattice and those in the α phase in a hexagonal close-packed lattice. Both are stacks of the same sort of close-packed planes. It is therefore extremely probable that the interface energy of the hydride in the α phase is considerably smaller than in the β phase in which the metal atoms form a completely different pattern (bcc). On theoretical grounds, therefore, it is to be expected that the formation of hydride nuclei will be much more difficult in the β phase than in the α phase. The same supposition has been made by Williams (l.c.) on the basis of experimental arguments.

Hydrogen-bearing α/β titanium alloys, like various hydrogen-containing steels, also exhibit the phenomenon of delayed brittle fracture (cf. Section 7.7). The incubation periods at ordinary temperatures are much longer than would be the case if the diffusion rate of hydrogen were the controlling factor. Nucleation of hydride may thus be supposed to be the critical step in this embrittlement process.

8.3. Phase separation in solutions of hydrogen in metals

If hydrogen is absorbed by a metal at constant temperature, then in general a solution of the gas in the metal will first be formed. In the case of titanium and zirconium, above a certain limiting concentration, in addition to the homogeneous solution, a hydride is formed with a crystal structure deviating from that of the metal (see Section 8.1). In the case of various other metals in which hydrogen dissolves exothermally, supersaturation does not lead to the formation of a hydride with different crystal structure, but to simple phase separation. This means that above a certain concentration a separation occurs into two solutions with the same crystal structure but different hydrogen concentrations.

The phase diagram in this case is of the simple type shown in Fig. 72. The hydrogen pressure increases along the curve with rising temperature. The miscibility gap $(\alpha + \beta)$ narrows with rising temperature. Above a critical temperature no phase separation occurs. The longest-known metal-gas system of this type is the palladium-hydrogen system, for which the critical temperature is about 300 °C. The corresponding values of the concentration and the hydrogen pressure are H/Pd $\cong$ 0.27 and $p(H_2) \cong$ 22 atm. The α

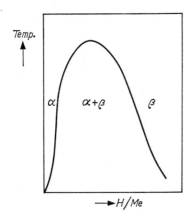

Fig. 72. Diagrammatic phase diagram of a metal-hydrogen system, in which phase separation occurs (examples Pd-H and Nb-H).

and β phases in this case are both *face-centred cubic*. As a result of their different hydrogen content, they differ in lattice parameter. With rising temperature the parameters of the co-existing α and β phases approach and become the same at the critical temperature. Fig. 73 gives these lattice parameters for the Pd-H system [1,2].

The dissociation pressures of palladium containing hydrogen at various concentrations and temperatures have been measured by many investigators [3-8]. Fig. 74 gives a number of pressure isotherms, taken from the cited

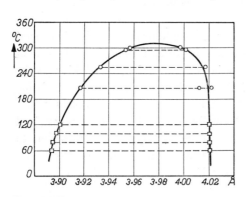

Fig. 73. Lattice parameters of the co-existent fcc phases α and β in the palladium-hydrogen system (squares after Owen and Williams, circles after Maeland and Gibb).

[1]) E. A. Owen and E. St. J. Williams, Proc. phys. Soc. **56**, 52 (1944)
[2]) A. J. Maeland and T. R. P. Gibb, J. phys. Chem. **65**, 1270 (1961)
[3]) C. Hoitsema, Z. phys. Chem. **17**, 1 (1895)
[4]) J. O. Linde and G. Borelius, Ann. Phys. **84**, 747 (1927)
[5]) H. Brüning and A. Sieverts, Z. physik. Chem. **A163**, 409 (1932-'33)
[6]) L. J. Gillespie and L. S. Galstaun, J. Amer. Chem. Soc. **58**, 2565 (1936)
[7]) P. L. Levine and K. E. Weale, Trans. Faraday Soc. **56**, 357 (1960)
[8]) More extensive literature survey in D. P. Smith, *Hydrogen in Metals*, University Press, Chicago 1948

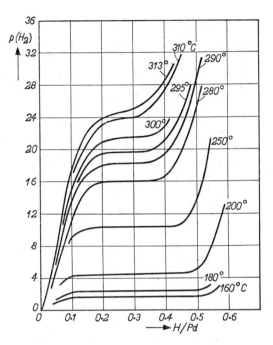

Fig. 74. Dissociation pressure (atm) of palladium-hydrogen alloys as a function of the H/Pd ratio for a number of temperatures (data of Sieverts, Gillespie and many others).

literature. In agreement with the phase rule they are horizontal in the region in which the two face-centred cubic phases (α and β) co-exist.

An example of a *body-centred cubic* metal in which the phase separation just discussed occurs, is niobium. Fig. 75 gives a log-log plot of equilibrium pressure against H/Nb ratio for a number of temperatures according to Albrecht, Goode and Mallett [1]). As the figure shows, the equilibria above 600 °C in the pressure range investigated obey the $\sqrt{p}$ law, according to which the hydrogen concentration in the metal is proportional to the square root of the equilibrium pressure of the molecular hydrogen. Therefore, at these temperatures and pressures the solutions can be regarded as dilute solutions of atomic hydrogen in niobium (see Section 7.2). At temperatures below 600 °C there are deviations from the $\sqrt{p}$ law, which are larger as the temperature is lower and the hydrogen pressure greater. We can then no longer speak of "dilute" solutions.

[1]) W. M. ALBRECHT, W. D. GOODE and M. W. MALLETT, J. Electrochem. Soc. **105**, 219 (1958) and **106**, 981 (1959)

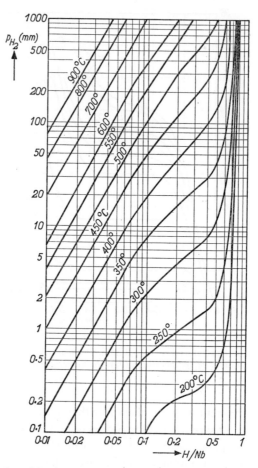

Fig. 75. Log-log plot of isotherms in the niobium-hydrogen system (Albrecht, Goode and Mallett).

It is a simple matter to derive the isobars in Fig. 76 from the isotherms in Fig. 75. None of the isotherms or isobars has a horizontal portion. At the temperatures and pressures at which the Nb-H system was studied there are thus no two-phase regions. The shape of the curves is such, however, that at lower temperatures and hydrogen pressures a miscibility gap can be expected with a critical temperature not far below 200 °C. Equilibrium with the gas phase is established so slowly below 200 °C that it is impossible to demonstrate such a phase separation by pressure measurements.

By means of X-ray diffraction measurements, however, the expected phase separation could easily be demonstrated by Albrecht et al. (l.c.). At room temperature it occurs in a wide region of hydrogen concentrations. This has been confirmed by Komjathy [1]) who, in accordance with other investigators [2,3]), finds that the β phase is slightly distorted. Albrecht et al. (l.c.) find by X-ray analysis at higher temperatures (20°-400 °C) that the critical point where the two bcc phases become identical is at about 140 °C and 0.3 H/Nb ratio. The equilibrium hydrogen pressure corresponding to this point was determined by extrapolation and is about 0.01 mm Hg. Fig. 72 not only applies to the Pd-H system, but also to the Nb-H system. The tantalum-hydrogen [4]) and vanadium-hydrogen [5]) systems are more complicated than the niobium-hydrogen system.

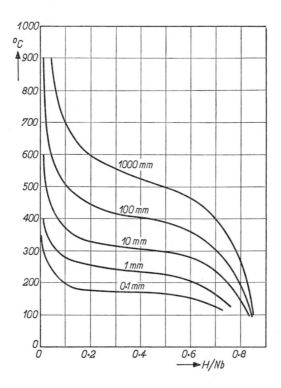

Fig. 76. Isobars corresponding to the isotherms in Fig. 75 (Albrecht, Goode and Mallett).

[1]) S. KOMJATHY, J. less-common Metals **2**, 466 (1960)
[2]) G. BRAUER and R. HERRMANN, Z. anorg. allg. Chem. **274**, 11 (1953)
[3]) C. WAINWRIGHT, A. J. COOK and B. E. HOPKINS, J. less-common Metals **6**, 362 (1964)
[4]) W. E. WALLACE and co-workers, Pure and appl. Chem. **2**, 281 (1961); J. chem. Phys. **35**, 2148 and 2156 (1961)
[5]) A. J. MAELAND, J. phys. Chem. **68**, 2197 (1964)

As was discussed in Section 7.3, the relative partial molar free enthalpy of hydrogen in a metal-hydrogen solution can be found directly from the equilibrium hydrogen pressure:

$$\Delta \mu_H = \Delta \bar{h}_H - T \Delta \bar{s}_H = \tfrac{1}{2} RT \ln p_{H_2}. \qquad (8.3.1)$$

In this equation $\Delta \bar{h}_H$ and $\Delta \bar{s}_H$ are the relative partial molar enthalpy and the relative partial molar entropy of hydrogen in the metal, i.e. the changes in enthalpy and entropy which occur when 0.5 mole H_2 at 1 atm is dissolved isothermally in an infinite quantity of the solid solution in question.

If $\Delta \mu_H$ for a Nb-H solution of constant concentration is plotted against the temperature, a straight line is obtained, the slope of which, according to equation (8.3.1), gives the relative partial molar entropy of hydrogen in the solution. This slope varies with the hydrogen content. The entropy values are thus invariant with temperature but vary with the hydrogen concentration of the solid solution. From the values found for $\Delta \mu_H$ and $\Delta \bar{s}_H$ values can be found for $\Delta \bar{h}_H$ by means of equation (8.3.1). In the case under consideration they are also invariant with temperature, but dependent on concentration.

Values for $\Delta \bar{h}_H$ can also be found in a slightly different manner. To do this, equation (8.3.1) is written in the form:

$$\ln p_{H_2} = \frac{2 \Delta \bar{h}_H}{RT} - \frac{2 \Delta \bar{s}_H}{R}. \qquad (8.3.2)$$

If the logarithm of $p(H_2)$ is plotted against $1/T$ at constant concentrations, straight lines are obtained, the slopes of which give the $\Delta \bar{h}$ values. Fig. 77 gives a number of these lines according to Komjathy (l.c.). Table 14 gives values of $\Delta \bar{h}$ and $\Delta \bar{s}$ by the same investigator.

TABLE 14

VALUES OF THE RELATIVE PARTIAL MOLAR ENTHALPY AND RELATIVE PARTIAL MOLAR
ENTROPY OF HYDROGEN IN NIOBIUM ACCORDING TO KOMJATHY.

H/Nb atomic ratio	$\Delta \bar{h}_H$ cal per gr.-atom H	$\Delta \bar{s}_H$ cal/degr. per gr.-atom H
0.05	−10680	−10.0
0.1	−11560	−12.5
0.2	−11920	−14.0
0.3	−13020	−16.0
0.4	−13440	−17.0
0.5	−14400	−19.0

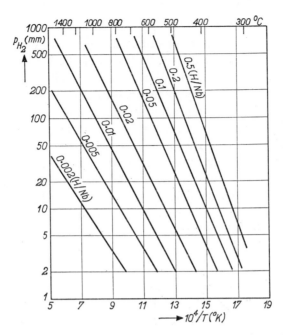

Fig. 77. Isocomposition curves (log p vs. $1/T$) in the niobium-hydrogen system (Komjathy).

The most striking thing about the table is that the absolute value of the relative partial molar enthalpy (the partial heat of solution) of hydrogen in the metal increases with increasing concentration. This is characteristic of the metal-gas systems (Pd-H, Nb-H, etc) in which phase separation occurs. In this type of system it is more advantageous from the point of view of energy if the hydrogen atoms are concentrated locally than if they are distributed evenly throughout the metal. At low temperatures, where the influence of entropy is small, separation into two phases must therefore be expected, one with a small concentration of hydrogen and one with the majority of the hydrogen in it. On the other hand, at high temperatures, where the influence of entropy is dominant, the occurrence of a single solution must be expected. As we have seen, this is exactly what is observed. The increase in the absolute value of the partial heat of solution of hydrogen with increasing hydrogen concentration thus gives the thermodynamic explanation for the phase separation observed. For a quantitative treatment of this problem on the basis of statistical-thermodynamic considerations, the reader is referred to the literature [1].

———————
[1] J. R. LACHER, Proc. Roy. Soc. A161, 525 (1937)

8.4. Metallic and saline hydrides

We have seen in the three preceding sections that hydrogen-rich phases occur in the systems Ti-H, Zr-H, Pd-H and Nb-H. These phases are often referred to as "metallic hydrides" because they exhibit a metallic electrical conductivity and also in their other properties resemble metals rather than salts. The same is true for the majority of the hydrides of the other transition metals. As indicated by the outlines in Fig. 78, nearly half the elements belong to the transition metals. They are characterized by an incomplete electron shell inside a completely or partially filled outer shell.

In their behaviour with respect to hydrogen, the group IV A metals titanium, zirconium, and hafnium show a great similarity. Hydrogen has a limited solubility in these metals and forms with all three a cubic and a tetragonal hydride with fluorite type structure. In the tetragonal hydride the

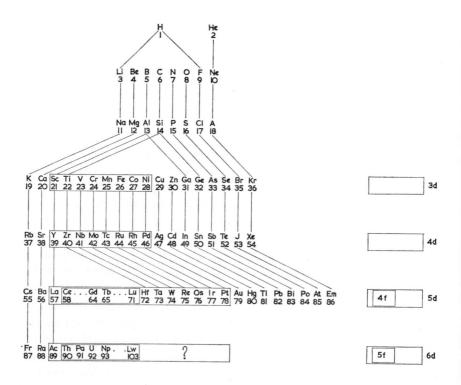

Fig. 78. Periodic system of the elements in which the transition metals are outlined. In the atoms of Sc, Ti, V, ... the 3d shell is only partly filled; in those of Y, Zr, Nb, ... this applies to the 4d shell, and so on.

hydrogen-to-metal ratio approaches two (see Section 8.1). Less similarity in their behaviour towards hydrogen is shown by the group VA metals vanadium, niobium and tantalum (see Section 8.3).

Most lanthanides (lanthanum, cerium, praseodymium, neodymium, . . ., lutetium) form two hydrides, the compositions of which correspond approximately to the formulae MeH_2 and MeH_3. Well-known exceptions are europium and ytterbium which form no trihydride. The heats of formation of the dihydrides lie around — 50 kcal/mole.

The interaction of the first four lanthanides (La, Ce, Pr and Nd) with hydrogen has been studied by Mulford and Holley [1]). When hydrogen is absorbed under equilibrium conditions by one of these metals and the terminal solubility is exceeded, the dihydride is formed. This can absorb a large quantity of additional hydrogen in homogeneous solution, to such an extent in fact that a region of continuous solid solution extends from about the composition MeH_2 to about MeH_3. One may therefore say that in the four systems under consideration there are not two hydrides, but only one with a large region of homogeneity, the composition being dependent on pressure and temperature. This is demonstrated by measurements of dissociation pressures at various temperatures and H/Me ratios. Figs. 79 and 80

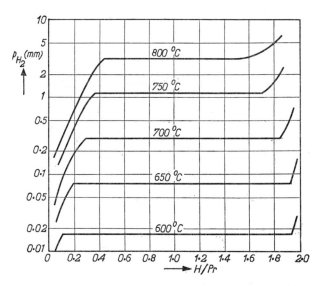

Fig. 79. Isotherms (log p versus H/Me) for the praseodymium-hydrogen system in the composition range Pr-PrH₂. The figures on the vertical axis give the pressures in mm Hg (Mulford and Holley).

[1]) R. N. R. MULFORD and C. E. HOLLEY, J. phys. Chem. 59, 1222 (1955)

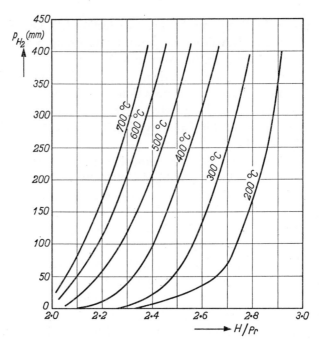

Fig. 80. Isotherms (p versus H/Me) for the praseodymium-hydrogen system in the composition range PrH$_2$-PrH$_3$. The figures on the vertical axis give the pressures in mm Hg (Mulford and Holley).

show a number of isotherms (dissociation pressure as a function of composition) for the praseodymium-hydrogen system. Fig. 79 relates to the composition range Pr-PrH$_2$ and Fig. 80 to the composition range PrH$_2$-PrH$_3$. In agreement with what has already been stated, application of the phase rule shows that in the latter range only one single homogeneous phase exists. The same is true for the regions to the left of the plateaus in Fig. 79, where we are dealing with solid solutions of hydrogen in the metal. The two solid phases co-exist in the composition ranges where a constant-pressure plateau occurs.

X-ray and neutron diffraction experiments [1]) showed that the MeH$_2$ phase (Me = La, Ce, Pr or Nd) has fluorite structure. As the hydrogen content increases from MeH$_2$ to MeH$_3$, the additional hydrogen atoms enter the octahedral interstices of the fluorite structure. This is accompanied by a *contraction* of the lattice.

---

[1]) C. E. HOLLEY et al., J. phys. Chem. **59**, 1226 (1955)

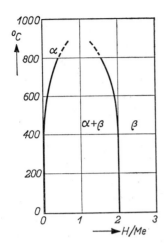

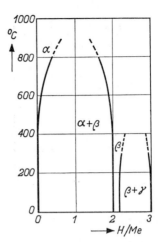

Fig. 81. Tentative solid phase diagram for the Me-H lanthanide systems where Me = La, Ce, Pr or Nd (Mulford).

Fig. 82. Tentative solid phase diagram for the Me-H lanthanide systems where Me = Sm, Gd, Dy or Er. The diagram is probably also valid for Me = Tb, Ho, Tm or Lu (Mulford).

In contrast to the trihydrides of La, Ce, Pr and Nd, those of the other lanthanides have a structure which deviates from that of their dihydrides. In Figs. 81 and 82, which give the tentative phase diagrams for a number of lanthanide metal-hydrogen systems according to Mulford, cited by Gibb [1]), this difference is expressed in the absence or presence of a second two-phase region $(\beta + \gamma)$.

The behaviour of the actinides (actinium, thorium, protactinium, uranium, etc.; see Fig. 78) with respect to hydrogen is less easily summarized than that of the lanthanides. Thorium forms two hydrides of which the limiting compositions at room temperature are ThH_2 and Th_4H_{15}. ThH_2 is isomorphous to ZrH_2: the metal atoms are arranged in a face-centred tetragonal lattice and the hydrogen atoms are situated in the tetrahedral interstices [2]). According to an investigation by Zachariasen [3]) the unit cell of the higher hydride contains 16 thorium atoms and 60 hydrogen atoms in beautiful agreement with the chemical formula Th_4H_{15}. Fig. 83 gives the equilibrium pressures of thorium-hydrogen alloys at various temperatures as a function of the H/Th ratio according to Nottorf, cited by Gibb (l.c.). The plateaus in the left-hand side of the figure correspond to the two-phase region $(Th + ThH_2)$, those in the right-hand side to the two-phase region $(ThH_2 + Th_4H_{15})$.

1) T. R. P. Gibb, Progr. inorg. Chem. 3, 315-509 (1962)
2) R. E. Rundle, C. G. Shull and E. O. Wollan, Acta Cryst. 5, 22 (1952)
3) W. H. Zachariasen, Acta Cryst. 6, 393 (1953)

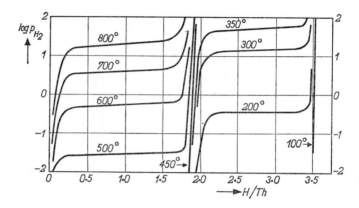

Fig. 83. Isotherms (logarithm of the equilibrium pressure in cm of mercury as a function of the H/Th ratio) for the thorium-hydrogen system. The left-hand plateaus give the dissociation pressures of "ThH$_2$", the right-hand plateaus those of "Th$_4$H$_{15}$" (Nottorf).

The solubility of hydrogen in thorium in equilibrium with the lower hydride increases rapidly with the temperature and, according to Peterson et al. [1,2], is about $2 \cdot 10^{-3}$ at% at room temperature, 1 at% at 300 °C and 26 at% at 800 °C. The hydride which is in equilibrium with these solutions has a H/Th ratio of about 2.00 at room temperature, 1.96 at 500 °C and 1.73 at 800 °C.

Uranium forms only one hydride with hydrogen, of approximate composition UH$_3$, in which the metal atoms form the cubic β tungsten structure. The solubility of hydrogen in the metal in equilibrium with this hydride is very small at ordinary temperatures [3]. Nearly all hydrogen which is present as an impurity in uranium (concentrations between one and five ppm are normal) precipitates as UH$_3$ at the grain boundaries, thereby diminishing the ductility of the metal [4,5,6]. Several investigators have shown that uranium hydride in equilibrium with uranium has a hydrogen content which is a little smaller than that corresponding to the formula UH$_3$. A report of a recent investigation and a survey of the literature on this subject can be found in an article by Besson and Chevallier [7].

The strongly electropositive alkali metals (Li, Na, K, Rb, Cs) and alkaline-

[1] D. T. PETERSON and D. G. WESTLAKE, Trans. AIME **215**, 444 (1959)
[2] D. T. PETERSON and J. REXER, J. less-common Metals **4**, 92 (1962)
[3] M. W. MALLETT and M. J. TRZECIAK, Trans. Amer. Soc. Met. **50**, 981 (1958)
[4] H. MOGARD and G. CABANE, Revue Métallurgie **51**, 617 (1954)
[5] H. R. GARDNER and J. W. RICHES, Trans. Amer. Soc. Met. **52**, 728 (1960)
[6] W. L. OWEN, Metallurgia **66**, 3 (July 1962)
[7] J. BESSON and J. CHEVALLIER, Comptes rendus **258**, 5888 (1964)

earth metals (Mg, Ca, Sr, Ba) form the saline or salt-like hydrides. They are all non-conductors of electricity and also in their other physical properties resemble salts. There is often a tendency to speak of covalent bonding in the case of the metallic hydrides and of ionic bonding in the case of the saline hydrides. In fact the bonding in both types of hydride is intermediate in nature. It can, however, be said that in most saline hydrides the ionic character of the bonding is dominant. This is most strongly evident in the alkali hydrides (LiH, NaH, etc.), which have the NaCl structure. Electrolysis of molten LiH yields hydrogen at the anode [1]). (The other alkali hydrides cannot be melted under normal hydrogen pressures without decomposing.) LiH and LiF form an uninterrupted series of solutions, not only in the liquid state, but also in the solid state above 300°-325 °C. This is demonstrated by the phase diagram in Fig. 84 [2]).

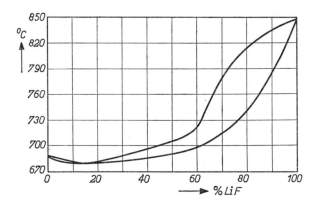

Fig. 84. Melting point phase diagram of the LiH-LiF system. The horizontal axis gives the LiF concentration in mole percent (Messer).

The isotherms which give the dissociation pressure as a function of the composition have the same shape for the saline metal-hydrogen systems as for the other metal-hydrogen systems in which only one hydride occurs. They also consist of three parts: (a) a sloping single-phase line corresponding to homogeneous solutions of the gas in the metal, (b) a plateau corresponding to the co-existence of hydrogen-saturated metal and hydride, (c) a sloping branch corresponding to a homogeneity region of the hydride. Fig. 85

[1]) K. MOERS, Z. anorg. allg. Chem. 113, 179 (1920)
[2]) C. E. MESSER, cited by GIBB, l.c. At room temperature LiH and LiF show very limited mutual solubility despite the closeness of the lattice parameters.

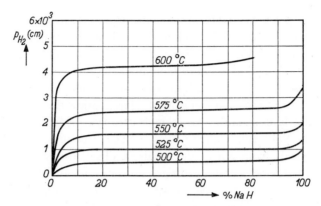

Fig. 85. Isotherms for the sodium-hydrogen system. The figures on the vertical axis give the pressure in cm of mercury. (Banus, McSharry and Sullivan).

demonstrates this with isotherms for the sodium-hydrogen system [1]), Fig. 86 with isotherms for the magnesium-hydrogen system [2]).

Since the advent of nuclear power, interest in metal hydrides has con-

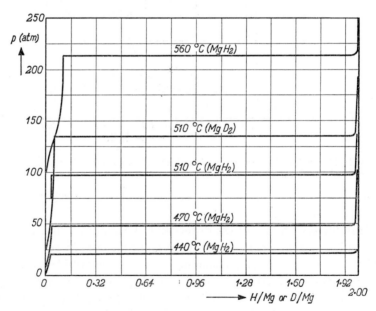

Fig. 86. Isotherms for the magnesium-hydrogen system. The figures on the vertical axis give the pressure in atm (Stampfer, Holley and Suttle).

---

[1]) M. D. BANUS, J. J. MCSHARRY and E. A. SULLIVAN, J. Amer. Chem. Soc. **77**, 2007 (1955)
[2]) J. F. STAMPFER, C. E. HOLLEY and J. F. SUTTLE, J. Amer. Chem. Soc. **82**, 3504 (1960)

siderably increased. The reasons are: (a) hydrogen is theoretically the most effective thermal neutron moderator, (b) the number of hydrogen atoms per unit volume in many metal hydrides is very large (larger than in liquid hydrogen), (c) many metal hydrides are still stable at relatively high temperatures. Of most interest at the present time is zirconium hydride because of its high hydrogen concentration, its relatively high thermal stability and the low cross-section of zirconium nuclei for neutron capture (0.18 barns).

More exhaustive information on metal hydrides than can be given in this book is to be found in review articles by Libowitz [1]) and Gibb [2]), the latter of which is the size of a book.

8.5. Solutions of oxygen and nitrogen in titanium, zirconium and hafnium

INTRODUCTION

Before 1925, zirconium and titanium were known as elements which are brittle by nature. After it had been shown that, by means of thermal decomposition of their tetraiodides, they could be obtained in the form of ductile rods [3,4]), it became clear that the brittleness was caused by impurities. Further investigation [5,6]) showed that zirconium and titanium can contain large quantities of oxygen and nitrogen in solid solution and that oxygen in particular is an impurity which is difficult to avoid. When present in solution this non-metallic element has a very detrimental effect on the mechanical properties of the two metals. Dissolved nitrogen has an even more unfavourable effect, but it is easier to prevent the absorption of this element than that of oxygen.

From a thermodynamic standpoint, one may speak either of solutions of oxygen and nitrogen or of solutions of oxide and nitride in the metal. A homogeneous solution of, say, oxygen in zirconium can be obtained by coating a rod of zirconium with a layer of ZrO_2 and subsequently heating it at a high temperature, but also by allowing it to react directly with gaseous oxygen at low pressure and high temperature. As has already been discussed

[1]) G. G. Libowitz, *The nature and properties of transition metal hydrides*, J. nuclear Mat. **2**, 1-22 (1960)

[2]) T. R. P. Gibb, *Primary solid hydrides*, Progr. inorg. Chem. **3**, 315-509 (1962)

[3]) J. H. de Boer and J. D. Fast, Z. anorg. allg. Chem. **153**, 1 (1926); **187**, 177 (1930)

[4]) J. D. Fast, Z. anorg. allg. Chem. **241**, 42 (1939)

[5]) J. H. de Boer and J. D. Fast, Rec. trav. chim. Pays-Bas **55**, 459 (1936); **59**, 161 (1940)

[6]) J. D. Fast, Metallwirtschaft **17**, 641 (1938)

in Section 6.2, it is somewhat more logical from an atomic standpoint to speak of solutions of oxygen and nitrogen than of solutions of oxide and nitride in a metal.

Oxygen and nitrogen as impurities in titanium and zirconium are so dangerous because, once they have been absorbed by the metal, they cannot be expelled again by heating in vacuum. This in contrast to hydrogen which, according to Section 8.2, also has an embrittling effect on titanium and zirconium, but which can be removed from the metal by heating.

In the well-known Kroll process [1]) titanium and zirconium with a low oxygen and nitrogen content are produced in the form of sponge by reduction of the tetrachlorides with magnesium. The process is characterized by the careful exclusion of air and water or water vapour during its critical steps. For consolidating the titanium and zirconium sponge into massive shapes without introducing oxygen and nitrogen, new melting techniques had to be developed of which consumable-electrode arc melting is of primary importance.

SOLUBILITY MEASUREMENTS

Solutions of oxygen and nitrogen in titanium and zirconium can be studied in various ways. The first proofs of the existence of exceptionally large solubilities were given by measurements of electrical resistance as a function of temperature before and after the absorption of known quantities of oxygen or nitrogen.

If the resistance of a wire of pure zirconium is measured as a function of temperature, a sudden drop is observed at 865 °C, the transition point from α Zr(hcp) to β Zr(bcc). This jump in the resistance amounts to 16.5% of the maximum value measured immediately before the transformation. After oxygen (or nitrogen) has been absorbed, the crystallographic transformation no longer occurs at one particular temperature, but is spread out over a wide temperature range (Fig. 87). Since this region lies wholly above the transition temperature of pure zirconium, the portion of the Zr-O phase diagram which relates to the crystallographic transformation is of the type illustrated in Fig. 27b. At each temperature in the transformation region an oxygen-rich α phase is in equilibrium with a β phase containing less oxygen. The Zr-N, Ti-O and Ti-N systems behave in an analogous manner. This is demonstrated

---

[1]) Extensive survey of literature in the books *The Metallurgy of Zirconium*, edited by B. Lustman and F. Kerze, McGraw-Hill, New York 1955, and *Titanium* by A. D. and M. K. McQuillan, Butterworth, London 1956.

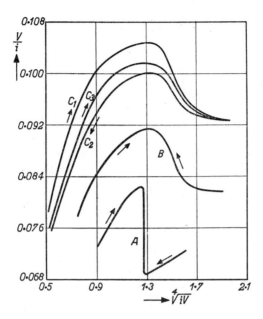

Fig. 87. Curve *A* gives the electrical resistance of a rod of pure zirconium as a function of the temperature. On transition from the hcp to the bcc modification (865 °C) there is a jump in resistance. After the absorption of 5 atom % *oxygen* and homogenization at high temperature, a transition region is observed which extends over hundreds of degrees centigrade (curve *B*). After the rod has also absorbed 5 atom % *nitrogen*, curves *C* are measured. As a measure of the temperature in this figure, the fourth root of the radiated energy has been chosen. This is not an accurate measure because only for a black body is the total radiation proportional to the fourth power of the absolute temperature. For many metals this power lies in between 4.5 and 5 over a wide range of temperatures (Curves taken from J. H. DE BOER and J. D. FAST, Rec. trav. chim. Pays-Bas **55**, 459, 1936).

by Figs. 88, 89, 90 and 91, which show the phase diagrams for these systems.

The most reliable data on the solubilities under discussion have been obtained by micrographic analysis and X-ray studies. Many investigators have observed that oxygen and nitrogen in solution cause an increase in the lattice parameters of hcp zirconium [1-4]) and titanium [5-9]). The densities

[1]) J. H. DE BOER and J. D. FAST, Rec. trav. chim. Pays-Bas **59**, 161 (1940)
[2]) R. M. TRECO, Trans. AIME **197**, 344 (1953)
[3]) R. F. DOMAGALA and D. J. MCPHERSON, Trans. AIME **200**, 238 (1954)
[4]) B. D. LICHTER, Trans. AIME **218**, 1015 (1960)
[5]) P. EHRLICH, Z. anorg. allg. Chem. **247**, 53 (1941)
[6]) H. T. CLARK, Trans. AIME **185**, 588 (1949)
[7]) E. S. BUMPS, H. D. KESSLER and M. HANSEN, Trans. Amer. Soc. Met. **45**, 1008 (1953)
[8]) P. EHRLICH, Z. anorg. allg. Chem. **259**, 1 (1949)
[9]) A. E. PALTY, H. MARGOLIN and J. P. NIELSEN, Trans. Amer. Soc. Met. **46**, 312 (1954)

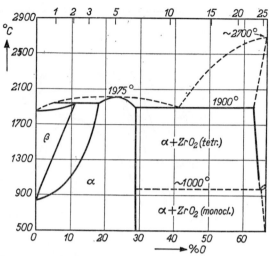

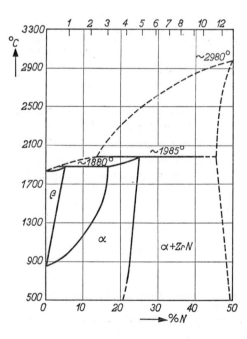

Fig. 88. Phase diagram of the Zr-ZrO₂ system according to DOMAGALA and MCPHERSON, Trans. AIME **200**, 238 (1954). The oxygen content is given on the lower horizontal axis in atom percent, on the upper one in weight percent. According to GEBHARDT, SEGHEZZI and DÜRRSCHNABEL, J. Nucl. Mat. **4**, 255 (1961), a few corrections should be made to this phase diagram.

Fig. 89. Phase diagram of the Zr-ZrN system according to DOMAGALA, MCPHERSON and HANSEN, Trans. AIME **206**, 98 (1956). The nitrogen content is given on the lower horizontal axis in atom percent, on the upper one in weight percent.

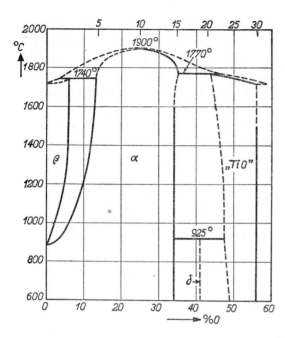

Fig. 90. Phase diagram of the titanium-oxygen system up to about 30 wt% oxygen according to BUMPS, KESSLER and HANSEN, Trans. Amer. Soc. Met. **45**, 1008 (1953). See also: SCHOFIELD and BACON, J. Inst. Metals **84**, 47 (1955). The oxygen content is given on the lower horizontal axis in atom percent, on the upper one in weight percent. The dotted part of the diagram has been slightly modified by the author in order to correct for an inconsistency.

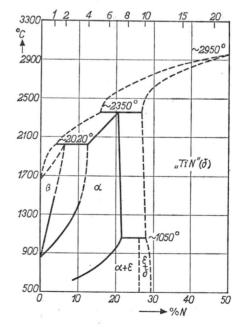

Fig. 91. Phase diagram of the titanium-nitrogen system up to 50 atom percent nitrogen according to PALTY, MARGOLIN and NIELSEN, Trans. Amer. Soc. Met. **46**, 312 (1954). The nitrogen content is given on the lower horizontal axis in atom percent, on the upper one in weight percent.

also increase when oxygen and nitrogen are dissolved. The measured densities agree quantitatively with those calculated from the lattice parameters, provided that it is assumed in the calculations that all the oxygen and nitrogen atoms are present in the interstices of the metal lattice, i.e. that no metal atoms have been replaced by O or N. Only the octahedral interstices can accommodate oxygen and nitrogen atoms.

To demonstrate how solubilities can be derived from measurements of lattice parameters, we shall show the results of measurements by Holmberg [1]) on Ti-N alloys quenched in water after a heat-treatment at 900 °C. Fig. 92 gives the lattice parameters of these alloys as functions of the nitrogen content. They are deduced from X-ray powder photographs. It follows immediately from the figure that the solubility limit of nitrogen in α Ti at 900 °C is close to the atomic ratio N/Ti = 0.20 (17 at% N) in satisfactory

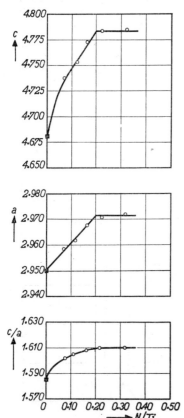

Fig. 92. Lattice parameters (in Å) of solid solutions of nitrogen in titanium heat-treated at 900 °C.

[1]) B. HOLMBERG, Acta Chem. Scand. 16, 1255 (1962)

agreement with the results of Ehrlich (l.c.). The X-ray patterns show no lines except those of a hcp atom arrangement, which is taken as evidence that the nitrogen atoms occupy the interstitial sites in a random way.

DISORDER AND ORDER IN SOLID SOLUTIONS OF OXYGEN IN Zr AND Ti

Solutions of *oxygen* in titanium and zirconium present a more complex picture than the solutions of nitrogen in Ti discussed above.

For zirconium-oxygen alloys heat-treated at 600 °C, Fig. 93 gives the lattice parameters and Fig. 94 the unit cell volume and the axial ratio as

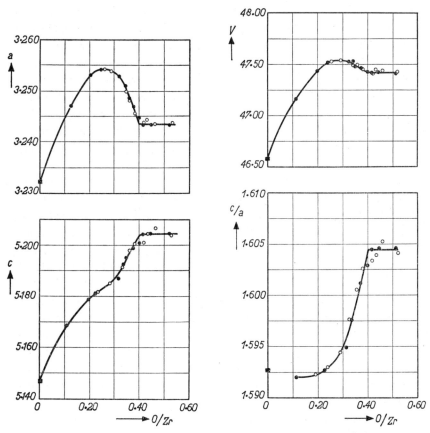

Fig. 93. Lattice parameters (in Å) of solid solutions of oxygen in zirconium heat-treated at 600 °C for one month (filled circles) and one week (open circles).

Fig. 94. Unit cell volume (in Å³) and axial ratio c/a of solid solutions of oxygen in zirconium heat-treated at 600 °C for one month (filled circles) and one week (open circles).

functions of the oxygen content [1]). According to these figures the solubility limit of oxygen in α Zr at 600 °C is close to the atomic ratio O/Zr = 0.40 (28.6 at% O), in good agreement with the results of Domagala and Mc-Pherson (l.c.). Above 400 °C the solubility is virtually independent of temperature. No reliable statements can be made about the solubility below 400 °C because the diffusion rate of oxygen in the metal is so small at low temperatures that equilibrium states are very difficult to attain.

In Fig. 93 it is striking that the length of the a axis passes through a pronounced maximum at the atomic ratio O/Zr ≅ 0.25 and that the rate of increase of the c axis changes markedly at the atomic ratio O/Zr ≅ 0.32. X-ray patterns have shown that in samples with oxygen contents below O/Zr = 0.34 the oxygen atoms are randomly distributed among the octahedral interstices. However, samples with atomic ratios O/Zr ≅ 0.35 show a completely ordered solid solution. The metal atoms in this structure have the same packing as in the pure metal but the oxygen atoms occupy a certain number of the octahedral interstices in an ordered way. The authors ascribe the ordered phase with the formula Zr_3O. It is isostructural with Ni_3N.

At oxygen contents greater than those corresponding to the composition Zr_3O the additional oxygen atoms are randomly distributed among the empty interstices of the Zr_3O structure. This region with a partly ordered solid solution exists up to the solubility limit O/Zr ≅ 0.40.

Also in titanium-oxygen solid solutions with high oxygen contents ordering phenomena occur [2]). There is, however, a considerable difference in the ways in which the oxygen atoms are arranged in the ordered solid solutions of the two metals. In the ordered Zr-O alloys the oxygen atoms are distributed in an ordered manner over all the layers of octahedral interstices extending normally to the c axis. In the ordered and partly ordered Ti-O solutions, on the other hand, the oxygen atoms reside exclusively in every second of these layers. In the saturated solution (atomic ratio O/Ti = 0.5) all the interstices in these planes are filled while those in the interleaving ones are empty. The metal atoms are not in the ideal positions of the hexagonal close-packed arrangement but somewhat displaced parallel to the c axis, away from the layers of oxygen atoms. This structure continues to exist even when the oxygen content is lowered. It then has randomly distributed vacancies in the oxygen arrangement.

Solutions with atomic ratio O/Ti = 0.33, after heat treatment at relatively low temperature (400 °C), exhibit a superstructure of the Ti_2O structure

[1]) B. HOLMBERG and T. DAGERHAMN, Acta Chem. Scand. **15**, 919 (1961)
[2]) B. HOLMBERG, Acta Chem. Scand. **16**, 1245 (1962)

discussed. In this superstructure the vacancies are present in an ordered distribution. They have a zig-zag arrangement in the direction of the c axis.

It will be clear to the crystallographer that the structures of the ordered Ti-O solutions are all of the anti-Cd(OH)$_2$ type.

In the zirconium-oxygen and titanium-oxygen phase diagrams (Figs. 88 and 90) the existence of the ordered phases just discussed is not shown. The required corrections can only be incorporated in the phase diagrams when more is known about the stability ranges of the ordered phases.

REDUCTION OF THE OXIDES WITH Mg, Ca OR Ba

The great solubility of oxygen in titanium and zirconium explains why it is impossible to prepare these metals in the pure state by reduction of their oxides with a metal which has a great affinity for oxygen, e.g. calcium.

It is seen in Table 1 (Chapter 3) that the absolute value of the free enthalpy of formation of 2 CaO is greater than that of TiO$_2$ or ZrO$_2$. Employing the well-known equation

$$\Delta G^0 = - RT \ln K_p$$

one can also express this by saying that the dissociation pressures of TiO$_2$ and ZrO$_2$ are larger than that of CaO. In this reasoning it is assumed, falsely, that when the oxides are reduced pure metal is immediately produced. If oxygen were insoluble in titanium and zirconium, one would therefore be able to produce these metals in the pure state by reduction of TiO$_2$ and ZrO$_2$ with calcium.

In reality, Ti and Zr are stabilized by dissolved oxygen, so that the dissociation pressures of the oxides in equilibrium with the saturated solutions are considerably larger than would be the case without this solubility. This seems favourable, but the existence of a solubility region also has the effect that the equilibrium pressure of oxygen above the metal does not suddenly drop to zero after the last vestiges of oxide have been removed. There is, on the contrary, a gradual decrease from the value corresponding to the solubility limit to the value zero for the pure metal. Reduction of the oxides with calcium therefore does not produce pure metal, but a metal-oxygen solution the dissociation pressure of which is equal to that of CaO at the prevailing temperature. The same concentration of oxygen in the metal is obtained if pure Ti or Zr is brought into contact with calcium oxide at the temperature concerned.

The dissociation pressure of the solution, in analogy to equation (8.3.1), is given by

$$\Delta\mu_O = \Delta\bar{h}_O - T\Delta\bar{s}_O = \tfrac{1}{2} RT \ln p_{O_2}, \tag{8.5.1}$$

where $\Delta\mu_O$, $\Delta\bar{h}_O$ and $\Delta\bar{s}_O$ are the relative values of the partial free enthalpy, the partial enthalpy and the partial entropy of a gramatom of oxygen in the metal, i.e. the changes which occur in the total values of these quantities when 0.5 mole O_2 at 1 atm dissolves isothermally in an infinitely large quantity of the given solid solution of oxygen in Ti or Zr.

The dissociation pressure of CaO, according to the equation

$$Ca + \tfrac{1}{2} O_2 \rightleftharpoons CaO,$$

is given by

$$\Delta G^0(CaO) = \tfrac{1}{2} RT \ln p_{O_2}, \tag{8.5.2}$$

where $\Delta G^0(CaO)$ is the standard value of the free enthalpy of formation of CaO at the temperature concerned. From (8.5.1) and (8.5.2) follows:

$$\Delta\mu_O = \Delta G^0(CaO). \tag{8.5.3}$$

This relationship also follows directly from the reaction equation

$$[O] + Ca \rightleftharpoons CaO, \tag{8.5.4}$$

where the symbol [O] indicates oxygen dissolved in Ti or Zr.

In contrast to $\Delta\mu_H$ values, $\Delta\mu_O$ values cannot be determined directly from equation (8.5.1) by pressure measurements since the equilibrium pressures are too small for this. With the aid of equation (8.5.3), however, it is possible to determine the relative partial molar free enthalpy of oxygen in the solution (i.e. $\Delta\mu_O$) by making use of the available tables for the free enthalpy of formation of CaO as a function of temperature [1]. The metal (Ti or Zr) is brought into equilibrium with an excess of calcium and calcium oxide at a certain high temperature and its oxygen content is subsequently determined by analytical means. This content corresponds to the value of the free enthalpy which is looked up in the tables mentioned. The procedure under discussion can be repeated for a number of other temperatures.

The method described was first employed by Allen and co-operators [2] to study the vanadium-oxygen system, later by Kubaschewski and Dench [3] to

---

[1] J. P. COUGHLIN, Bulletin 542, Bureau of Mines: Contributions to the data on theoretical metallurgy, XII. Heats and free energies of formation of inorganic oxides, Washington 1954

[2] N. P. ALLEN, O. KUBASCHEWSKI and O. VON GOLDBECK, J. Electrochem. Soc. **98**, 417 (1951)

[3] O. KUBASCHEWSKI and W. A. DENCH, J. Inst. Met. **82**, 87 (1953/54); **84**, 440 (1955/56)

study the titanium-oxygen and zirconium-oxygen systems. As reducing agent they used, not only calcium, but also magnesium and barium. For any temperature at which the three reduction equilibria of the type (8.5.4) are established quickly enough, it is possible in this way to obtain the values of $\Delta\mu_O$ for three different oxygen concentrations in the metal. Employing the relationship

$$\Delta\mu_O = RT \ln a_O$$

the three corresponding oxygen activities are easily found.

According to the experiments of Kubaschewski and Dench (l.c.) the reduction of titanium oxide with Ca, Mg or Ba at 1000 °C yields Ti with the following equilibrium oxygen concentrations:

Reduction by:	Ca	Mg	Ba
Oxygen in Ti, wt%	0.07	2.3	6.6

CALORIMETRIC DETERMINATION OF THERMODYNAMIC QUANTITIES

Making use of equation (8.5.1) the chemical potential of oxygen in solution in Ti or Zr can also be found in another way, viz. by determining the partial heat of solution $\Delta\bar{h}_O$ and the partial entropy of solution $\Delta\bar{s}_O$. This method was chosen by a number of research workers of the U.S. Bureau of Mines [1].

Heats of formation of a number of titanium-oxygen interstitial solutions were determined by combustion calorimetry. We shall describe the method, using as example a solution containing 99.76 wt% Ti and 0.24 wt% O, i.e. a solution with atom fractions (mole fractions) Ti = 0.9928 and O = 0.0072. By the combustion of 1 gram-atom of solution to 0.9928 mole of rutile a quantity of heat was liberated, which after correction to a constant pressure and to 298 °K is given by

$$\Delta H_{298}(\text{sol.}) = -223,186 \text{ cal.}$$

The combustion of 0.9928 gr.-atom of pure titanium to form 0.9928 mole of rutile gives

$$\Delta H_{298}(\text{TiO}_2) = -224,135 \text{ cal.}$$

---

[1] A. D. MAH, K. K. KELLEY, et al., Report of Investigations 5316, Bureau of Mines: Thermodynamic properties of titanium-oxygen solutions and compounds, Washington 1957

It is clear that the difference between these two results,

$$\Delta H_{298} = -950 \text{ cal,}$$

is the heat of formation of the interstitial solution from the elements.

In this way the heats of formation were determined for nine interstitial solutions with atom fractions of oxygen in the range 0 to 0.056. Per gr.-atom of solution they are all given by

$$\Delta H_{298}(\text{form.}) = -138,800 \, x_O,$$

where x_O is the atom fraction of oxygen. In the concentration range under discussion, the *partial* heat of solution is thus independent of the concentration:

$$\Delta \bar{h}_O(298 \text{ °K}) = -138,800 \text{ cal.}$$

It will be noted that this heat of solution refers to 1 gr.-atom of oxygen and that its absolute value is greater than that of the heat of formation of TiO_2 per gr.-atom of oxygen.

The entropies of solution at absolute zero temperature were calculated by Mah and Kelley (l.c.) by statistical means (cf. Chapter 1). In the calculations it was assumed that the oxygen in the dilute solutions resides entirely in the octahedral interstices and that it occupies them randomly [1]. The entropy increments for the solutions between 0° and 298 °K were calculated from the results of low-temperature heat capacity measurements. Fig. 95 gives these results for pure titanium and for two interstitial solutions.

From the values found for the partial heat of solution and the partial entropy of solution, the values of $\Delta \mu_O$ at 298 °K were calculated by means of equation (8.5.1). The calculation of $\Delta \mu_O$ values at temperatures above the transition point $a \rightarrow \beta$ (885 °C) is more complicated because Ti-O solutions at these high temperatures, dependent on their oxygen content, may be in the a region, the β region or the two-phase region. It is therefore necessary to make use, not only of the results of heat-capacity measurements at high temperatures, but also of the position of the phase boundaries in the phase diagram. Furthermore, until they have been measured, certain plausible assumptions must be made about the heat capacities of β solutions.

The values of $\Delta \mu_O$ which have been calculated for 1000 °C by Mah and Kelley in the manner indicated, agree not unsatisfactorily with the values

---

[1] The calculations of MAH and KELLEY are not entirely correct because they assume that the number of octahedral interstices is equal to half the number of titanium atoms, while in fact the numbers are equal.

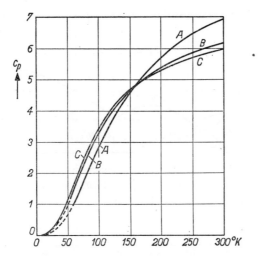

Fig. 95. Low-temperature heat capacities c_p (cal/deg. mole) of titanium and titanium-oxygen solid solutions. Curve A: $TiO_{0.334}$. Curve B: $TiO_{0.062}$. Curve C: pure Ti. (Mah, Kelley, Gellert, King and O'Brien)

deduced by Kubaschewski and Dench from their reduction and oxidation experiments with calcium and calcium oxide.

THE HAFNIUM-OXYGEN SYSTEM

The solubilities, just discussed, of oxygen and nitrogen in titanium and zirconium are exceptionally large. Of all the other metals only hafnium, which is closely related to Ti and Zr, has a comparable capacity to dissolve the two gases. Fig. 96 shows the hafnium-oxygen phase diagram according to Rudy and Stecher [1]. The solubility of oxygen in the α phase (hcp hafnium), according to these investigators, is practically independent of the temperature and is 20.5 atom percent. This value had already been found for one particular temperature (600 °C) by Dagerhamn [2].

Comparison with the foregoing shows that the solubility of oxygen in the group IV A metals decreases with increasing atomic number (Ti: 33.3 at%, Zr: 28.6 at%, Hf: 20.5 at%). Ordered arrangements of the interstitial oxygen atoms, such as were found in the solutions of oxygen in titanium and zirconium, do not seem to occur in the hafnium-oxygen system.

[1] E. RUDY and P. STECHER, J. less-common Metals 5, 78 (1963)
[2] T. DAGERHAMN, Acta Chem. Scand. 15, 214 (1961)

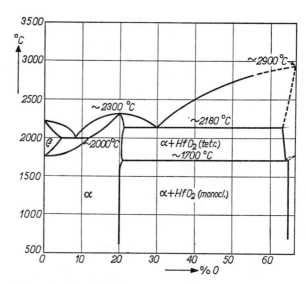

Fig. 96. Phase diagram of the Hf-HfO₂ system according to RUDY and STECHER, J. less-common Metals **5**, 78 (1963). The horizontal axis gives the oxygen content in atom percent.

8.6. Solutions of oxygen and nitrogen in vanadium, niobium and tantalum

The metals which, after Ti, Zr and Hf, can absorb the next largest quantities of oxygen and nitrogen in solid solution are the body-centred cubic group VA metals: vanadium, niobium and tantalum. Contrary to what is true for Ti, Zr and Hf, oxygen and nitrogen can be expelled almost completely from Nb and Ta by heating at a high temperature in a good vacuum.

SOLUBILITY OF OXYGEN IN V, Nb AND Ta

Our knowledge of the vanadium-oxygen, niobium-oxygen and tantalum-oxygen systems is still very incomplete, also with respect to the solubility of oxygen in the metallic phase. Very roughly, it can be said that this solubility for all three metals is of the order of magnitude of 1 atom percent at 600 °C, while it is several atom percent at 1600 °C. At equal temperatures it is of the same order of magnitude as the solubility of oxygen in the high temperature modifications (β modifications) of Ti, Zr and Hf, which are also body-centred cubic (cf. Figs. 88, 90 and 96).

For the *vanadium-oxygen* system [1]) the literature gives two partial phase

----

[1]) Recent review: J. STRINGER, J. less-common Metals **8**, 1 (1965).

diagrams [1,2]), which differ considerably from one another. According to Seybolt and Sumsion the solubility of oxygen in solid vanadium is 3.2 at% and virtually independent of temperature. According to Rostoker and Yamamoto, on the other hand, the solubility decreases with decreasing temperature from about 3 at% at 1840 °C to less than 0.8 at% at 500 °C. Both investigations indicate that when the solubility limit is exceeded a body-centred tetragonal oxide is formed, which has approximately the composition V_5O when in equilibrium with the saturated solution.

Very little is known about the phase diagram of the *tantalum-oxygen* system. If tantalum is oxidized at various oxygen pressures and temperatures, X-ray and metallographic studies [3-7]) show not only dissolution of oxygen in the metal, but also formation of one or more of the following oxides: TaO_y, TaO_z, TaO, $\beta\,Ta_2O_5$ and $\alpha\,Ta_2O_5$. For the first two only the crystal structures are known, but not the exact values of y and z. Probably TaO_y, TaO_z and TaO are metastable oxides which are not included in the equilibrium diagram Ta-O.

Gebhardt and Seghezzi [8]) find the values given in the table below for the solubility of oxygen in tantalum. They state that these are solubilities of

Temp. (°C)	700	900	1100	1300	1650
Solubility (at% O)	1.5	2.2	3.1	4.2	6.4

oxygen in equilibrium with a suboxide of unknown composition. However, with the investigations of Kofstad (l.c.) in mind, it must be assumed that at the three highest temperatures the metal was in equilibrium with Ta_2O_5 [9]). This assumption is supported by a plot of $\log c$ against $1/T$ (Fig. 97). The figure shows that the points corresponding to the three highest temperatures lie on a straight line, while those corresponding to 700° and 900 °C lie considerably higher. This pattern can be explained by assuming that the metal above 1000 °C was in equilibrium with the stable oxide phase (Ta_2O_5), and below

[1]) A. U. SEYBOLT and H. T. SUMSION, Trans. AIME **197**, 292 (1953)
[2]) W. ROSTOKER and A. S. YAMAMOTO, Trans. Amer. Soc. Metals **47**, 1002 (1955)
[3]) N. NORMAN, J. less-common Metals **4**, 52 (1962)
[4]) N. NORMAN, P. KOFSTAD and O. J. KRUDTAA, J. less-common Metals **4**, 124 (1962)
[5]) G. BRAUER, H. MÜLLER and G. KÜHNER, J. less-common Metals **4**, 533 (1962)
[6]) P. KOFSTAD, J. Inst. Metals **91**, 209 (1962/63); J. Electrochem. Soc. **110**, 491 (1963); J. less-common Metals **5**, 158 (1963)
[7]) M. G. COWGILL and J. STRINGER, J. Inst. Metals **91**, 220 (1962/63)
[8]) E. GEBHARDT and H. D. SEGHEZZI, Z. Metallkde **50**, 521 (1959)
[9]) See also: L. B. DUBROVSKAYA et al., Shurnal neorg. Khimii **9**, 1182 (1964); in Russian

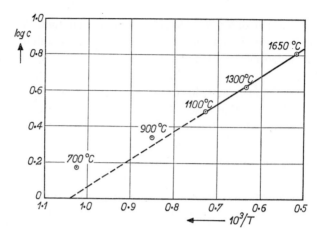

Fig. 97. Solubility of oxygen in tantalum at five different temperatures according to GEBHARDT and SEGHEZZI, Z. Metallkde **50**, 521 (1959). Log c (at % O) has been plotted against $1/T$ (°K).

1000 °C with metastable oxides. (The solubility of a metastable compound is always greater than that of the stable compound).

According to more recent work by Vaughan, Stewart and Schwartz [1] the solubility of oxygen in tantalum in equilibrium with Ta_2O_5 is 3.65 at % at 1500 °C, 2.95 at % at 1000 °C and 2.5 at % at 500 °C. The points obtained in this case by plotting log c against $1/T$ do not lie on a straight line, so that we are inclined to attach more importance to the high-temperature values of Gebhardt and Seghezzi (straight line in Fig. 97).

The *niobium-oxygen* phase diagram (Fig. 98) has been determined by Elliott [2]. According to his research the solid solubility of oxygen in niobium varies from 1.4 at % at 500 °C to 4.0 at % at the eutectic temperature (1915 °C). Seybolt [3] found values which do not agree with this, viz. 1.4 at % at 775 °C and 5.5 at % at 1100 °C. The most reliable values for the solubility of oxygen in niobium are probably those of Bryant [4] and of Gebhardt and Rothenbacher [5]), which only differ slightly from each other. The results of the measurements by Bryant, which were carried out in the temperature range 700°-1550 °C, are given in Fig. 99 and by the equation

$$c = 59.7 \exp\left(-8600/RT\right) \quad \text{at \% O} \tag{8.6.1}$$

[1] D. A. VAUGHAN, O. M. STEWART and C. M. SCHWARTZ, Trans. AIME **221**, 937 (1961)
[2] R. P. ELLIOTT, Trans. Amer. Soc. Metals **52**, 990 (1960)
[3] A. U. SEYBOLT, Trans. AIME **200**, 774 (1954)
[4] R. T. BRYANT, J. less-common Metals **4**, 62 (1962)
[5] E. GEBHARDT and R. ROTHENBACHER, Z. Metallkde **54**, 443 and 623 (1963)

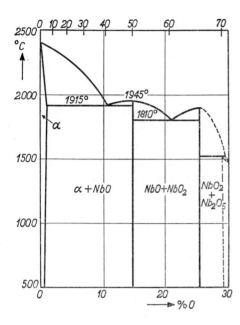

Fig. 98. Phase diagram of the niobium oxygen system according to ELLIOTT, Trans. Amer. Soc. Met. **52**, 990 (1960). The oxygen content is given on the lower horizontal axis in weight percent, on the upper one in atom percent.

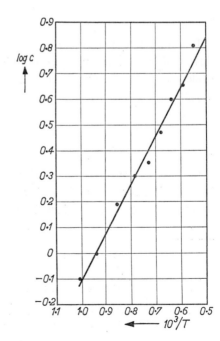

Fig. 99. Solubility of oxygen in niobium according to BRYANT, J. less-common Metals **4**, 62 (1962). Log c (at % O) has been plotted versus $1/T$ (°K).

or, in another form:

$$\log c \,(\text{at}\% \, \text{O}) = -\frac{1880}{T} + 1.776. \qquad (8.6.2)$$

The heat of solution associated with the dissolution of oxygen from the oxide phase is therefore 8.6 kcal/gr.-atom oxygen (7.7 kcal according to Gebhardt and Rothenbacher). The oxide phase existing in equilibrium with the saturated solid solution is NbO (see Fig. 98). According to the equation the solubility of oxygen in niobium in equilibrium with NbO is 0.70 at% at 700 °C and 5.2 at% at 1500 °C.

Gebhardt and Rothenbacher obtained supersaturated solutions of oxygen in niobium by quenching saturated solutions from 1460 °C. In agreement with the phase diagram by Elliott (Fig. 98), a precipitate of NbO forms in these supersaturated solutions when they are heated at a temperature between 800° and 1000 °C. If they are heated at a temperature below 700 °C, not only NbO precipitates, but also one or two suboxides. These are probably identical with oxides found by Hurlen [1] and Norman and collaborators [2,3] after the oxidation of niobium below 600 °C. They are often referred to as NbO_z and NbO_x. According to Brauer and collaborators [4] NbO_x has the composition Nb_6O. The occurrence of the two suboxides is not necessarily incompatible with Elliott's equilibrium diagram (Fig. 98), since they are probably metastable phases.

In addition to confirming the existence of niobium suboxides Van Landuyt [5], using electron microscopy and electron diffraction techniques, found evidence of the occurrence of ordering phenomena in solid solutions of oxygen in niobium.

SOLUBILITY OF NITROGEN IN V, Nb AND Ta

Little is known of the solubility of *nitrogen in vanadium* [6]. During an investigation of the vanadium-nitrogen system Hahn [7] found three phases: the metallic phase (α phase) in which very little nitrogen dissolves, a hexagonal

[1] T. HURLEN, J. Inst. Metals **89**, 273 (1960/61)
[2] N. NORMAN, J. less-common Metals **4**, 52 (1962)
[3] N. NORMAN, P. KOFSTAD and O. J. KRUDTAA, J. less-common Metals **4**, 124 (1962)
[4] G. BRAUER, H. MÜLLER and G. KÜHNER, J. less-common Metals **4**, 533 (1962)
[5] J. VAN LANDUYT, Physica status solidi **6**, 957 (1964)
[6] W. ROSTOKER and A. YAMAMOTO, Trans. Amer. Soc. Metals **46**, 1136 (1954)
[7] H. HAHN, Z. anorg. allg. Chem. **258**, 58 (1949)

subnitride (β phase) with a homogeneity range $VN_{0.37}$-$VN_{0.43}$ and a cubic nitride (δ phase) with a homogeneity range $VN_{0.71}$-$VN_{1.00}$. Brauer and Schnell [1]) find virtually the same homogeneity range for the cubic nitride as Hahn, but a greater range ($VN_{0.37}$-$VN_{0.49}$) for the hexagonal subnitride.

According to an X-ray investigation by Vaughan, Stewart and Schwartz [2]) the solubility of *nitrogen in tantalum* in equilibrium with a subnitride of unknown composition is about as large as that of oxygen in tantalum in equilibrium with Ta_2O_5. At 1500 °C, they find a solubility of 3.70 at% N, at 1000 °C of 2.75 at% N and at 500 °C of 1.8 at% N.

Gebhardt, Seghezzi and Fromm [3]) find considerably larger solubilities. Following an earlier investigation [4]) they made an extensive study of the tantalum-nitrogen system by measuring equilibrium nitrogen pressures, by electrical resistance measurements and by microscopic and X-ray investigation. Fig. 100 gives a log-log plot of the equilibrium pressures of Ta-N alloys as a function of the nitrogen content at various temperatures. The horizontal portions of these isotherms correspond to the three-phase equilibria between

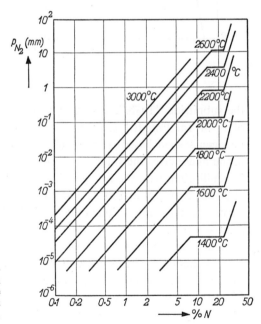

Fig. 100. Logarithmic plot of isotherms in the tantalum-nitrogen system. The horizontal axis gives the nitrogen content in atom percent (Gebhardt, Seghezzi and Fromm).

[1]) G. BRAUER and W. D. SCHNELL, J. less-common Metals **6**, 326 (1964)
[2]) D. A. VAUGHAN, O. M. STEWART and C. M. SCHWARTZ, Trans. AIME **221**, 937 (1961)
[3]) E. GEBHARDT, H. D. SEGHEZZI and E. FROMM, Z. Metallkde **52**, 464 (1961)
[4]) E. GEBHARDT, H. D. SEGHEZZI and W. DÜRRSCHNABEL, Z. Metallkde **49**, 577 (1958)

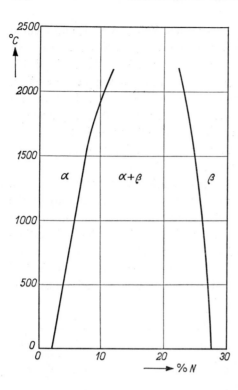

Fig. 101. Part of the phase diagram of the tantalum-nitrogen system, in which the nitrogen content is given in atom percent. (Gebhardt, Seghezzi and Fromm).

the subnitride "Ta$_2$N", gaseous nitrogen and the saturated solid solution of N in Ta. Fig. 101 gives part of the phase diagram for the tantalum-nitrogen system based on the various measurements. The left-hand curve in the figure gives the solubility of nitrogen in body-centred cubic tantalum in equilibrium with the hexagonal subnitride "Ta$_2$N" (solubility 2 at% N at 20 °C, 5 at% N at 700 °C and 13 at% N at 2300 °C). The right-hand curve gives the composition of the subnitride in equilibrium with the saturated solid solution.

Measurements by Bunn and Wert [1]) of the internal friction of nitrogen-bearing tantalum wires show that the solubilities at low temperatures (350° to 550 °C) are much smaller than the values given by the left-hand curve of Fig. 101. The values determined by Gebhardt et al. (l.c.) at high temperatures (above 1500 °C) combine very well with those found by Bunn and Wert at low temperatures: a straight line is obtained by plotting log c against $1/T$.

[1]) P. BUNN and C. A. WERT, Trans. AIME **230**, 936 (1964)

Over the entire temperature range 350°-2300 °C the concentration of nitrogen in equilibrium with a nitride precipitate is given by the equation

$$c = 35 \exp \left(-5400/RT\right) \quad \text{at}\%$$ (8.6.3)

or, in another form:

$$\log c \, (\text{at}\%) = -\frac{1180}{T} + 1.544.$$ (8.6.4)

According to this equation the solubility of nitrogen in tantalum in equilibrium with a nitride (presumably "Ta_2N") is 0.45 at% at 350 °C and 12.2 at% at 2300 °C.

The nice agreement between the measurements of Gebhardt et al. (l.c.) at high temperatures and those of Bunn and Wert (l.c.) at low temperatures is remarkable on the ground of an investigation by Seraphim, Stemple and Novick [1]. These investigators find that the random solutions of nitrogen in tantalum which contain more than 0.1 at% N are not stable at low temperatures. Heat treatment at, say, 400 °C leads to the formation of an ordered phase with an approximate composition $Ta_{27}N$ in the interstitial solid solutions. The bcc superlattice cell of this phase contains 27 expanded cells of the random solid solution. This is in agreement with the description already given earlier by Schönberg [2] of an ordered phase with approximately the composition $TaN_{0.05}$. He found also the nitrides $TaN_{0.40-0.45}$ ("Ta_2N"), $TaN_{0.8-0.9}$ and TaN. The existence of all these nitrides has been confirmed by a new investigation [3].

The ordering in the interstitial solutions mentioned above indicates strong interaction between the interstitial atoms in tantalum. This interaction also manifests itself in the internal friction of solid solutions of oxygen and nitrogen in tantalum [4].

The solubility of *nitrogen in niobium* has been studied extensively by Cost and Wert [5]. As Gebhardt, Seghezzi and Fromm (l.c.) did for solutions of nitrogen in tantalum, so Cost and Wert measured the equilibrium N_2 pressures of solutions of nitrogen in niobium. Fig. 102 gives a log-log plot of the

[1] D. P. SERAPHIM, N. R. STEMPLE and D. T. NOVICK, J. appl. Phys. **33**, 136 (1962)
[2] N. SCHÖNBERG, Acta Chem. Scand. **8**, 199 (1954)
[3] K. ÖSTHAGEN and P. KOFSTAD, J. less-common Metals **5**, 7 (1963)
[4] R. W. POWERS and M. V. DOYLE, Acta Met. **4**, 233 (1956); J. appl. Phys. **28**, 255 (1957); J. Metals **9**, 1287 (1957); Acta Met. **6**, 643 (1958); J. appl. Phys. **30**, 514 (1959); Trans. AIME **215**, 655 (1959)
[5] J. R. COST and C. A. WERT, Acta Met. **11**, 231 (1963)

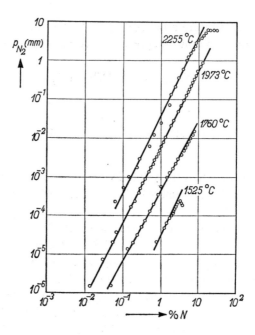

Fig. 102. Logarithmic plot of isotherms in the niobium-nitrogen system. The horizontal axis gives the nitrogen content in atom percent (Cost and Wert).

nitrogen pressure as a function of the composition for four different temperatures. The straight lines drawn through the data points have the slope 2 which shows that the $\sqrt{p}$ law holds quite well over the complete temperature range of the measurements and over more than six orders of magnitude of pressure. The highest value for the nitrogen concentration at each temperature corresponds roughly to the solubility limit (terminal solubility) at that temperature.

Fig. 103 gives on a logarithmic scale the nitrogen pressure as a function of $1/T$ (T = absolute temperature) for a number of constant concentrations. The slopes of these lines give the values of the relative partial molar enthalpy or partial molar heat of solution, $\Delta \bar{h}_N$, of nitrogen in niobium (cf. Section 8.3). From the fact that the lines are virtually straight and mutually parallel, it follows that the partial molar heat of solution is virtually independent of concentration and temperature. According to the figure it is -46.0 ± 2.0 kcal per gr.-atom of nitrogen.

Since $\Delta \bar{h}_N$ is constant and the $\sqrt{p}$ law valid over a wide range of concentrations, Cost and Wert conclude that the solid solutions of nitrogen in

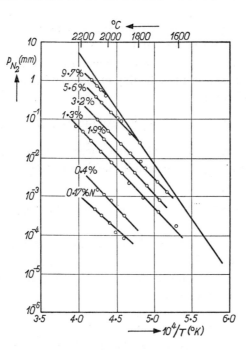

Fig. 103. Isocomposition curves (log p versus $1/T$) in the niobium-nitrogen system. The line farthest right in the figure, having a markedly different slope from the others, gives the dissociation pressure of "Nb_2N" in equilibrium with the saturated solution of nitrogen in Nb (Cost and Wert).

niobium behave as nearly ideal solutions, at least at temperatures above 1500 °C (see Figs. 102 and 103).

The line in Fig. 103 which has a slope differing markedly from that of the other lines gives the decomposition pressure of the nitride "Nb_2N", the compound which forms on the niobium after the solubility limit is exceeded. It has the same crystal structure as "Ta_2N" (metal atoms hexagonally close-packed) and, like this compound, posesses a wide homogeneity range extending from 28.6 to 33.3 at % N [1,2]). The "Nb_2N" line in the figure relates to the nitrogen-poor boundary of this phase. It gives the equilibrium pressures in the case where the compound is in equilibrium with the saturated solid solution of nitrogen in niobium. The solubility lines intersect the "Nb_2N" line with falling temperature. We are here dealing with the case which, in the discussion of the zirconium-hydrogen system, has already been schematically

[1]) G. BRAUER and J. JANDER, Z. anorg. allg. Chem **270**, 160 (1952)
[2]) N SCHÖNBERG, Acta Chem. Scand. **8**, 208 (1954)

represented in Fig. 66 (note that $1/T$ increases from right to left in Fig. 66 and from left to right in Fig. 103).

From the measurements at high temperatures (Figs. 102 and 103) the following is deduced for the solubility of nitrogen in niobium in equilibrium with gaseous nitrogen:

$$c(N_2) = 6.2 \times 10^{-4} \sqrt{p_{N_2}} \exp (46,000/RT) \quad \text{at} \% \, N, \qquad (8.6.5)$$

where p is in mm Hg. From the same measurements follows for the high temperature solubility of nitrogen in niobium in equilibrium with "Nb_2N" (terminal solubility):

$$c(\text{"}Nb_2N\text{"}) = 720 \exp (-20,000/RT) \quad \text{at} \% \, N. \qquad (8.6.6)$$

Results of recent measurements by Gebhardt, Fromm and Jakob [1]) on the solubility of nitrogen in niobium in the temperature range 1600°-2200 °C are in good agreement with the results of Cost and Wert (l.c.) given by the equations (8.6.5) and (8.6.6).

For temperatures below 1150 °C the solubility of N in Nb in equilibrium with "Nb_2N" was deduced by Cost and Wert (l.c.) from internal friction measurements. The circles in Fig. 104 give the results of pressure measurements at high temperatures (equation 8.6.6)), while the triangles represent the results of the internal friction measurements. According to the figure the solubility of nitrogen in niobium in equilibrium with "Nb_2N" at low temperatures is much greater than would be expected on the basis of equation (8.6.6). The results would seem to indicate that the heat of solution of Nb_2N in Nb decreases continuously from a high temperature value of 20 kcal per gr.-atom of nitrogen to a low temperature value of 4.6 kcal, which had already been found earlier by Ang and Wert [2]). It is not yet possible to give a satisfactory explanation of the remarkable shape of the solubility curve in Fig. 104. The question arises whether the saturated solution of nitrogen in niobium is in equilibrium with one and the same nitride over the whole temperature range.

Lack of sufficient data makes it impossible to construct a complete phase-diagram for the niobium-nitrogen system. A tentative diagram can be found in an article by Brauer and Esselborn [3]). In addition to the compound "Nb_2N" mentioned, there exist other niobium nitrides richer in nitrogen.

[1]) E. GEBHARDT, E. FROMM and D. JAKOB, Z. Metallkde 55, 423 (1964)
[2]) C. Y. ANG and C. A. WERT, Trans. AIME 197, 1032 (1953)
[3]) G. BRAUER and R. ESSELBORN, Z. anorg. allg. Chem. 309, 151 (1961)

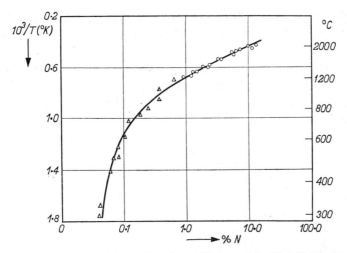

Fig. 104. Solubility of nitrogen in niobium in equilibrium with "Nb₂N". The horizontal axis gives the nitrogen content in atom percent. The circles give the results of pressure measurements at high temperatures (Figs. 102 and 103), the triangles give the results of measurements of internal friction (Cost and Wert).

They have compositions which are given approximately by the formulae Nb_4N_3 and NbN. The latter compound occurs in at least two different crystal modifications [1]).

8.7. Oxygen and nitrogen in metals

SOLUBILITIES

The elements in groups IV A and V A (Ti, Zr, Hf, V, Nb, Ta) are the only ones in the periodic system which in the solid state are able to dissolve large quantities of both oxygen and nitrogen. Elements in which only one of the two gases has a large solid solubility (say greater than 1 at%) are, in fact, even rarer.

Nitrogen, as already discussed in Section 7.9, has a large solubility in the fcc (γ) phase of *iron.* According to Fig. 52 this solubility (in equilibrium with Fe_4N) is 2.4 wt% or about 9 at% N at the eutectoid temperature, i.e. at 590 °C. The available data also show that nitrogen has a large solubility in

[1]) G. Brauer and H. Kirner, Z. anorg. allg. Chem. **328**, 34 (1964)

the high temperature modifications of *manganese*, the largest (about 20 at %) in the fcc (γ) modification. Fig. 105 gives part of the tentative phase diagram of the manganese-nitrogen system at a total pressure of 1 atm. It was compiled by Hansen, Zwicker and Motz [1,2]. In this system four compounds occur, which can be roughly indicated by the formulae Mn_4N, Mn_5N_2, M_3N_2 and Mn_6N_5 [3]. The last two have such a high nitrogen content that they lie outside the scope of Fig. 105.

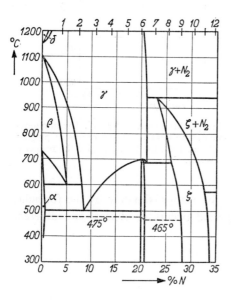

Fig. 105. Tentative isobaric manganese-nitrogen phase diagram according to Hansen, Zwicker and Motz. The lower horizontal axis gives the nitrogen content in atom percent and the upper one in weight percent.

Nitrogen also has a fairly large solubility in solid *chromium*. Fig. 106 gives the solubility of nitrogen in chromium in equilibrium with Cr_2N according to Caplan, Fraser and Burr [4]. The results are in satisfactory agreement with those of Seybolt and Oriani [5]. The solubility is given by the equation

$$\log c \, (\text{wt} \% \text{ N}) = -\frac{7250}{T} + 3.86. \qquad (8.7.1)$$

This equation gives a solubility of 3×10^{-6} wt % N (10^{-5} at % N) at 500 °C and 0.59 wt % N (2.2 at % N) at 1500 °C. The solubility of *oxygen*

[1] M. HANSEN, *Constitution of Binary Alloys*, McGraw-Hill, New York 1958 (p. 936)
[2] U. ZWICKER, Z. Metallkde **42**, 274 (1951)
[3] F. LIHL, P. ETTMAYER and A. KUTZELNIGG, Z. Metallkde **53**, 715 (1962)
[4] D. CAPLAN, M. J. FRASER and A. A. BURR, *Ductile Chromium and its Alloys*, Amer. Soc. Met., Cleveland 1957 (p. 196)
[5] A. U. SEYBOLT and R. A. ORIANI, Trans. AIME **206**, 556 (1956)

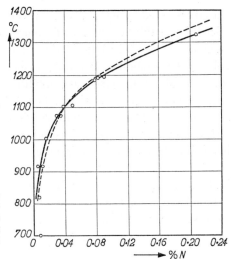

Fig. 106. Solubility of nitrogen in chromium in equilibrium with Cr₂N according to Caplan, Fraser and Burr. The dotted line gives the solubility according to Seybolt and Oriani. The nitrogen content is given in weight percent.

in chromium (in equilibrium with Cr_2O_3) is very much smaller than that of nitrogen in equilibrium with Cr_2N. An analogous statement is valid for the solubilities of oxygen and nitrogen in iron and manganese.

Oxygen does not dissolve in large quantities in any solid metal outside groups IV A and V A. Even the well-known and much-discussed solubility of oxygen in *silver* is very small by comparison. According to Steacie and Johnson [1] it exhibits a minimum in solid silver at 400 °C. More recent investigations have shown that this minimum is not the result of an exceptional variation of the solubility with temperature but of the simultaneous adsorption of oxygen [2] or of reactions of oxygen with trace impurities, such as copper [3]. The solubility itself exhibits no minimum.

The investigators mentioned [1-3] did not determine the terminal solubilities of oxygen in silver in equilibrium with Ag_2O, but the quantities of gas which are absorbed by the metal below the solubility limit at certain temperatures and pressures. Since the dissociation pressure of Ag_2O is known as a function of temperature and the quantity dissolved is proportional to the square root of the oxygen pressure, it is possible to calculate the terminal solubilities from the measured solubilities. In Table 15 we find in the first column the temperature in degrees centigrade and in the second column the solubilities found by Eichenauer and Müller (l.c.), in cm^3 O_2 (at 0 °C and

[1] E. W. R. STEACIE and F. M. G. JOHNSON, Proc. Roy. Soc. A 112, 542 (1926)
[2] W. EICHENAUER and G. MÜLLER, Z. Metallkde 53, 321 (1962)
[3] H. H. PODGURSKI and F. N. DAVIS, Trans. AIME 230, 731 (1964)

1 atm) absorbed by 100 g silver at an O_2 pressure of 1 atm. To this we have added: in the third column the decomposition pressures of Ag_2O, calculated from $\log p = -2859/T + 6.2853$ [1]) (p in atm, T in °K); in the fourth column the solubilities at these pressures (i.e. the terminal solubilities); in the fifth column the terminal solubilities in atom percent. These have only been given up to 500 °C, since Allen [2]) found that it is impossible for *solid* silver to be in equilibrium with Ag_2O above 507 °C.

The solubility of oxygen in solid silver at an oxygen pressure of 1 atm just below the melting point is about 5 cm^3 O_2 (at 0 °C and 1 atm) per 100 g Ag. Just above the melting point this solubility under the same conditions is about 230 cm^3 per 100 g Ag [3]). The great difference in solubility of oxygen in liquid and in solid silver causes the well-known phenomenon of "spitting" when silver melted in air solidifies.

TABLE 15

SOLUBILITY OF OXYGEN IN SOLID SILVER

Temp. °C	Solubility at 1 atm $cm^3 O_2$/100 g Ag	Decomposition pressure of Ag_2O atm	Terminal solubility $cm^3 O_2$/100 g Ag	Terminal solubility at % O
200	0.0023	1.74	0.0030	3×10^{-5}
300	0.021	19.8	0.093	9×10^{-4}
400	0.098	109	1.02	1×10^{-2}
500	0.308	386	6.05	6×10^{-2}

MECHANICAL PROPERTIES

A dissolved gas (oxygen, nitrogen or hydrogen) makes a metal harder and more brittle. The influence on the mechanical properties is even greater if the same quantity of gas, after precipitation from a supersaturated solid solution, is present in the metal in the form of an extremely finely divided compound. For nitrogen in bcc iron this is clearly illustrated by Fig. 63. The well-known and very injurious influence of nitrogen on the mechanical properties of *chromium* [4]) is probably to be explained in this way. We have already seen (Fig. 106) that nitrogen at high temperatures has a relatively

[1]) F. G. KEYES and H. HARA, J. Amer. Chem. Soc. **44**, 479 (1922). The equation differs slightly from that calculated in Section 4.1, which relates to temperatures below 200 °C.
[2]) N. P. ALLEN, J. Inst. Metals **44**, 317 (1932). According to this author Ag and Ag_2O form a eutectic mixture at 507 °C and 414 atm.
[3]) A. SIEVERTS and J. HAGENACKER, Z. physik. Chem. **68**, 115 (1909)
[4]) H. L. WAIN, F. HENDERSON and S. T. M. JOHNSTONE, J. Inst. Metals **83**, 133 (1954/55)

large solubility in chromium, while at low temperatures it has a very small solubility. Even small quantities of dissolved nitrogen will therefore show a tendency to precipitate at low temperatures in the form of very finely divided Cr_2N.

The influence of *dissolved* oxygen and nitrogen on the hardness of a metal is demonstrated by Fig. 107 with titanium as an example [1]. Relatively small quantities of oxygen and nitrogen have a great strengthening effect and ultimately a very marked embrittling effect on titanium at room temperature. The same is true for the other metals in groups IV A and V A.

The effect of oxygen and nitrogen on the mechanical properties of a metal is only slight when they occur in the metal in the form of relatively coarse oxide and nitride particles. For example, *thorium* may contain many at% of oxygen without being brittle. Due to its very small solubility oxygen is almost entirely present as a dispersion of ThO_2 in a ductile thorium matrix [2,3]. According to Peterson [4] the solubility of oxygen in thorium in equilibrium with ThO_2 is only 35 ppm (0.0035 wt% or 0.05 at%) at 1000 °C and

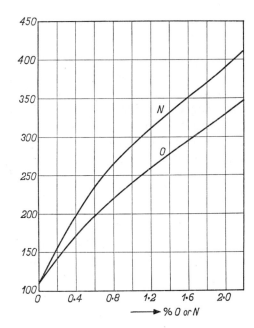

Fig. 107. Vickers hardness H of iodide titanium as a function of the concentration of dissolved oxygen or nitrogen in at% (Jaffee, Ogden and Maykuth).

[1] R. I. JAFFEE, H. R. OGDEN and D. J. MAYKUTH, Trans. AIME **188**, 1261 (1950)
[2] J. D. FAST, Metallwirtschaft **17**, 641 (1938)
[3] M. D. SMITH and R. W. K. HONEYCOMBE, J. Nucl. Mat. **1**, 345 (1959)
[4] D. T. PETERSON, Trans. AIME **221**, 924 (1961)

90 ppm (0.009 wt% or 0.13 at%) at 1200 °C. By compacting and sinter-ing one can even obtain ductile thorium from a powder of which the grains are coated with a thick skin of oxide. This is explained by the fact that during sintering the skins collect into separate oxide inclusions [1]).

Not only thorium, but also metals such as scandium, yttrium, the lanthani-des and uranium, notwithstanding their great affinity for oxygen, can be prepared in ductile form with comparative ease as a result of their small dissolving power for oxygen. All these metals can even be separated from their oxides by fusion.

There are other metals, however, of which the mechanical properties can be influenced most unfavourably by oxygen although their power to dissolve this gas is negligibly small. An example of such a metal is *iron* [2]). An oxygen content of the order of magnitude of 0.01 wt% is sufficient to make pure iron brittle. In measurements of the impact strength of normalized material as a function of temperature, the influence of oxygen is clearly demonstrated by a shift of the ductile to brittle transition temperature to higher temper-atures. Microscopic examination of the impact specimens after fracture shows that this shift is due to grain boundary embrittlement. In the specimens tested below the transition range the fractures are entirely of the cleavage type in the case of pure iron, whereas they are partly intergranular in the case of iron containing about 0.01 wt% oxygen.

The reduction of the grain boundary fracture strength of iron by the presence of oxygen is the more striking because *carbon* may be present at the grain boundaries without weakening them. We have tried to explain this by assuming the oxide to be present along the grain boundaries in the form of almost uninterrupted skins, whereas the carbide forms separate particles at the boundaries. The hypothetical oxide skins would have to be extremely thin, because even with the aid of the optical and electron microscopes we were unable to make them visible. One only sees globular oxide particles which show no preference for the crystal boundaries but are randomly distributed in the metal. Thus, in order to understand the weakening of the grain boundaries one should, perhaps, think in terms of intergranular adsorption or segregation of oxygen rather than of intergranular oxide skins.

The grain boundary embrittlement of iron by oxygen has been confirmed and thoroughly studied in the National Physical Laboratory in England [3]).

---

[1]) J. D. FAST, Philips techn. Review **3**, 345 (1938)
[2]) J. D. FAST, *International Foundry Congress, Amsterdam 1949*, De Hofstad, The Hague (p. 171)
[3]) W. P. REES and B. E. HOPKINS, J. Iron Steel Inst. **169**, 157 (1951) and **172**, 403 (1952)

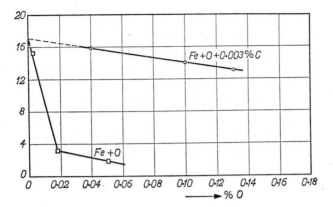

Fig. 108. Impact strength (kgm/cm²) at 20 °C of iron as a function of the oxygen content (wt %). Traces of carbon greatly reduce the deteriorating effect of oxygen on the impact strength (Fast).

Curiously enough, it turns out that traces of carbon (a few thousandths percent) are sufficient to reduce considerably the injurious effect of even large quantities of oxygen in iron [1]. This is demonstrated by Fig. 108 which gives the impact strength of iron at 20 °C as a function of the oxygen content. The small amounts of carbon were introduced in the Fe-O alloys by melting them under certain CO pressures. It seems reasonable to suppose that the traces of carbon alter the distribution of the intergranular oxygen in such a way that a finely divided oxide precipitate forms at the grain boundaries.

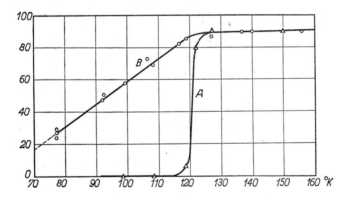

Fig. 109. Percentage of reduction in area in tensile tests as a function of temperature. Curve *A* relates to iron containing 0.02 wt % oxygen, curve *B* to iron containing only 0.003 wt % oxygen but also to iron containing 0.02 wt % O plus 0.004 wt % C (Gibbons).

---

[1] J. D. FAST, *International Foundry Congress, Amsterdam 1949*, De Hofstad, The Hague (p. 171)

The remarkable effect of carbon traces on the mechanical properties of iron containing oxygen is also clearly demonstrated by tensile tests of Gibbons [1]). In Fig. 109 curve *A* gives the ductility, as measured by the percentage of reduction in area at fracture, for wire specimens containing 0.02 wt% oxygen. At about 125 °K the ductility suddenly drops to zero. Iron containing only 0.003% oxygen shows no sudden loss in ductility (curve *B*). If the iron containing 0.02% oxygen is carburized to a value of 0.004% C, it no longer exhibits the sudden loss of ductility at 125 °K (curve *A*), but behaves like the iron containing only 0.003% oxygen (curve *B*).

8.8. Simultaneous presence of oxygen and hydrogen in metals

Hall, Martin and Rees [2]) and Singh and Parr [3]) have measured the solubility of hydrogen in zirconium-oxygen alloys. Fig. 110 gives a number of isotherms (quantity of hydrogen absorbed as a function of the H_2 pressure)

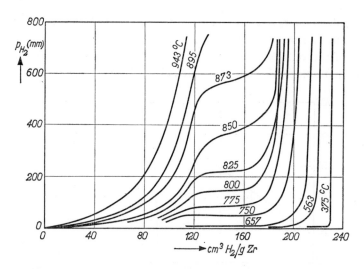

Fig. 110. Isotherms which give the quantity of hydrogen taken up by a zirconium-oxygen solid solution with atomic ratio $O/Zr = 0.023$ (Hall, Martin and Rees).

[1]) D. F. GIBBONS, Trans. AIME **197**, 1245 (1953)
[2]) M. N. A. HALL, S. L. H. MARTIN and A. L. G. REES, Trans. Faraday Soc. **41**, 306 (1945) and **50**, 343 (1954)
[3]) K. P. SINGH and J. GORDON PARR, Trans. Faraday Soc. **59**, 2248 (1963)

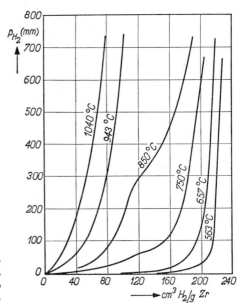

Fig. 111. Isotherms giving the quantity of hydrogen taken up by a zirconium-oxygen solid solution with atomic ratio O/Zr = 0.058 (Hall, Martin and Rees).

for an alloy with atomic ratio O/Zr = 0.023 and Fig. 111 for an alloy with atomic ratio O/Zr = 0.058.

The solubility of hydrogen in titanium-oxygen solid solutions has been studied by Hepworth and Schuhmann [1]). Nearly all their measurements refer to a temperature of 800 °C. Fig. 112 gives an isothermal section of the Ti-O-H system at 800 °C. The curves marked A to L are isoactivity curves for hydrogen, i.e. each of these curves corresponds to a particular hydrogen pressure above the ternary alloys and thus to a particular activity of hydrogen in the metal. The dotted portion of each isoactivity curve is a tie line in the two-phase region ($\alpha + \beta$). A corresponds to an H_2 pressure of about 5 mm, L to an H_2 pressure of about 170 mm.

Fig. 112 shows that in the binary system Ti-H at 800 °C the hcp α region extends to 3.5 at% H. From 3.5 to 6.8 at% H the oxygen-free alloys are two-phase and above 6.8 at% H the bcc β phase is stable (cf. the analogous binary diagram Zr-H in Fig. 65). With increasing oxygen content a considerable widening of the two-phase region occurs. The trend of the isoactivity curves shows that the solubility of hydrogen in both α and β titanium decreases with increasing oxygen content. According to Brown and Hardie [2]),

[1]) M. T. HEPWORTH and R. SCHUHMANN, Trans. AIME **224**, 928 (1962)
[2]) A. BROWN and D. HARDIE, J. Nucl. Mat. **4**, 110 (1961)

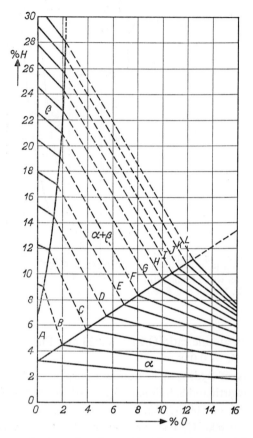

Fig. 112. Ti-O-H system isothermal section at 800 °C showing isoactivity curves, *A* to *L,* for hydrogen. The hydrogen and oxygen contents ar given in atom percent (Hepworth and Schuhmann).

the same is not true for the ternary system Zr-O-H. They conclude from their experiments that oxygen *increases* the solubility of hydrogen in *α* zirconium and *decreases* it in *β* zirconium.

Hepworth and Schuhmann (l.c.) deduce the partial molar heats of solution for hydrogen in Ti-O solid solutions (Table 16) from a number of measurements with varying temperature. The values relating to pure titanium are by McQuillan [1].

---

[1] A. D. McQuillan, Proc. Roy. Soc. A **240**, 309 (1950)

TABLE 16

PARTIAL MOLAR HEATS OF SOLUTION OF HYDROGEN IN Ti-O SOLID SOLUTIONS CORRE
SPONDING TO THE REACTION $H_2 \rightarrow 2$ [H] α or β

O/Ti	$\Delta \bar{h}$, cal per mole H_2	
	α	β
0	-21600	-27830
0.0051	-21480 ± 1500	-27000 ± 1500
0.0281	-21810 ± 1500	—
0.0378	-21110 ± 1500	-25700 ± 1500
0.0938	—	-23900 ± 1500
0.1229	-21070 ± 1500	-24600 ± 1500

Hepworth and Schuhmann (l.c.) use their measured hydrogen activities in the ternary alloys Ti-O-H in an elegant manner to calculate
the activities of oxygen and titanium in the binary system Ti-O. First
they calculate the activities of these elements in the ternary alloys by
means of the Gibbs-Duhem equation as it has been extended to ternary systems [1-3]. They then determine the activities in the binary
system Ti-O by extrapolation.

The gas phase in equilibrium with the ternary alloys discussed (Zr-O-H
and Ti-O-H) consists of practically pure hydrogen. It is true that in principle
it also contains traces of water vapour according to the equation

$$[O] + 2\,[H] \rightleftharpoons H_2O, \tag{8.8.1}$$

but the partial pressure of the water vapour is negligibly small. It is absolutely impossible to reduce Zr-O and Ti-O solid solutions by means of hydrogen. This impossibility does not apply to solutions of oxygen in metals which
have a much smaller affinity for oxygen than zirconium and titanium. In the
ternary system, one is then dealing with equilibria in which the gas phase
contains water vapour in addition to hydrogen. In such cases the oxygen
content of the metal can be lowered with the help of hydrogen and the hydrogen content with oxygen.

As an example of this we shall discuss the technically important case of
molten iron which, according to Sections 6.2 and 7.4, can contain both oxygen
and hydrogen in solution. For small H and O concentrations the reaction
constant corresponding to equation (8.8.1) can be written approximately in
the form

[1]) C. WAGNER, *Thermodynamics of Alloys*, Addison-Wesley Press, 1952.
[2]) L. S. DARKEN, J. Amer. Chem. Soc. **72**, 2909 (1950)
[3]) R. SCHUHMANN, Acta Met. **3**, 219 (1955)

$$K = \frac{p(H_2O)}{[\% \, H]^2 \cdot [\% \, O]}, \qquad (8.8.2)$$

where the activities of H and O have been replaced by their percentages. For liquid iron and a temperature of 1540 °C the value of K is about 10^6, if $p(H_2O)$ is expressed in atmospheres and $[\% \, H]$ and $[\% \, O]$ in weight percent. If the liquid iron is in contact with a gaseous atmosphere in which $p(H_2O)$ has a constant value, then the product $[\% \, H]^2 \cdot [\% \, O]$ will also have a constant value. It follows that molten iron with a low constant oxygen activity will absorb much more hydrogen than liquid iron with a high oxygen activity, if it comes into contact with an atmosphere containing water vapour.

This point has played an important part in the development of electrodes for arc welding of steel. The older types of electrode had a coating containing much iron oxide, so that the deposited metal contained much oxygen. As a result the mechanical properties of the weld were relatively poor. By only using very stable oxides (CaO, MgO, Al_2O_3, TiO_2, SiO_2) and by adding reducing metal powders to the coating, it gradually became possible to make coatings which introduced less and less oxygen into the deposited metal. At first, however, it was not realized that the ever-improving solution to the oxygen-problem was bound to lead to an increased absorption of hydrogen, unless the water content of the coating was drastically reduced at the same time. Instead of the difficulties caused by oxygen, one therefore obtained difficulties caused by hydrogen [1]. Eventually success was achieved in making "low-hydrogen" welding electrodes which deposit metal with both a low oxygen content and a low hydrogen content.

Also in the melting and casting of *copper* the interaction between oxygen and hydrogen in the liquid metal plays an important part. In order to prevent hydrogen difficulties, especially the occurrence of porosity on solidification, copper is usually melted in an oxidizing atmosphere or under an oxidizing slag. The molten copper then contains relatively large amounts of oxygen in solution and as a result absorbs but little hydrogen from the ever-present water vapour.

Immediately before casting the copper is deoxidized by means of phosphorus, calcium or magnesium. The quantity of hydrogen which is present in the metal after oxidizing melting and deoxidation is too small to produce appreciable porosity. If deoxidation before casting is omitted, then in the

[1] J. D. FAST, Welding J. **32**, 516 (1953); Schweissen und Schneiden **9**, 512 (1957)

last stages of the solidification process porosity may arise due to the forma-tion of water vapour bubbles. These are formed as soon as the product $[\% \text{ H}]^2 \cdot [\% \text{ O}]$ in the remaining liquid exceeds a definite value.

Copper-tin and copper-aluminium alloys cannot contain oxygen in solu-tion, since all the oxygen absorbed precipitates as a second phase (tin oxide or aluminium oxide). An oxidizing atmosphere or slag is therefore no protection in this case against the penetration of hydrogen. In order to prevent damage by hydrogen in these alloys they are often flushed with an inert gas before casting (cf. Section 6.6).

CHAPTER 9

SOLUTIONS OF GASES IN ALLOYS

9.1. Activity coefficients in multi-component solutions

In technology one frequently has to deal with solutions of gases in binary, ternary or multi-component alloys. Think, for example, of solutions of oxygen, nitrogen or hydrogen in liquid steel. Even the simplest type of steel always contains several solutes (C, Mn, Si, S, P), while "alloy steel" also contains larger or smaller percentages of other added metals.

The solubility of a gas in a metal can be either increased or decreased by the presence of other solutes. In many cases it is possible to forecast the sign of the change in solubility on the grounds of the respective atomic bond energies. Let us take as an example the solubility of nitrogen in liquid iron containing another solute. If the bond energy nitrogen-solute is greater than the bond energies nitrogen-iron and solute-iron, then in general the solubility of nitrogen will be increased. An example of a solute having this effect is manganese. Conversely, if the bond energy solute-iron is greater than the other two, the solubility of nitrogen will generally be decreased. An example of a solute of this kind is sulphur. On the average, each sulphur atom will have a greater number of iron atoms surrounding it than that corresponding to a random distribution of the three atomic species. This rejection of the nitrogen atoms from groups of sulphur and iron atoms implies the decrease of solubility mentioned above. An analogous argument can be made to explain the increase in solubility of nitrogen as a result of the presence of dissolved manganese.

In the thermodynamic treatment the effects under discussion are manifested in a change in the activity and activity coefficient of the dissolved gas under the influence of the other solutes. It should be remembered that the thermodynamic behaviour of a solute is not determined by its concentration c but by its activity a, the latter being expressed as the product of an activity coefficient f and the concentration c:

$$a = fc. \tag{9.1.1}$$

The concentration c may be expressed in any convenient unit. In the solutions in question, where one component preponderates, the activities are generally so defined that they become equal to the concentrations in the infinitely dilute solutions (see Section 6.4). In the following, we too shall make use of this definition where necessary. It implies that the activity coefficients are equal to unity in the infinitely dilute solutions.

In a dilute solution of $(n-1)$ components, 2, 3, 4, . . ., n, in a solvent metal 1 the activity coefficient of any solute may usually be expressed as a product of $(n-1)$ factors [1]. For example, the activity coefficient of component 2 may be expressed as

$$f_2 = f_2^{(2)} f_2^{(3)} f_2^{(4)} \ldots f_2^{(n)}, \tag{9.1.2}$$

where $f_2^{(2)}$ is the activity coefficient of 2 in the binary solution (1,2) containing the same mole fraction of 2 as the multi-component solution and where the other factors are correction factors which account for the presence of the components 3, 4, . . ., n.

Using a Taylor series expansion for the logarithm of the activity coefficient, Wagner [2] derived the following equivalent expression:

$$\ln f_2 = x_2 \frac{\partial \ln f_2}{\partial x_2} + x_3 \frac{\partial \ln f_2}{\partial x_3} + \ldots + x_n \frac{\partial \ln f_2}{\partial x_n}, \tag{9.1.3}$$

where x_2, x_3, . . ., x_n are the mole fractions of the solutes, and where the derivatives are to be taken for the limiting case of zero concentration of all solutes, which implies that they are constants. Terms of the Taylor series involving second and higher derivatives have been neglected.

Frequently (9.1.3) is written in the form

$$\ln f_2 = x_2 \varepsilon_2^{(2)} + x_3 \varepsilon_2^{(3)} + \ldots + x_n \varepsilon_2^{(n)}, \tag{9.1.4}$$

where the symbols $\varepsilon_2^{(2)}$, $\varepsilon_2^{(3)}$, . . ., $\varepsilon_2^{(n)}$ are defined as

$$\varepsilon_2^{(2)} = \frac{\partial \ln f_2}{\partial x_2} ; \varepsilon_2^{(3)} = \frac{\partial \ln f_2}{\partial x_3} ; \ldots ; \varepsilon_2^{(n)} = \frac{\partial \ln f_2}{\partial x_n}. \tag{9.1.5}$$

There exists a simple relationship (see Wagner, l.c.) between the effect of component i on f_j and that of component j on f_i:

$$\varepsilon_j^{(i)} = \varepsilon_i^{(j)}. \tag{9.1.6}$$

If the concentrations are expressed in weight percent instead of mole fractions

[1] J. CHIPMAN, J. Iron Steel Inst. **180**, 97 (1955).
[2] C. WAGNER, *Thermodynamics of Alloys* (p. 51), Addison-Wesley Press, Cambridge, U.S.A., 1952.

and if common logarithms are used instead of natural logarithms, then equation (9.1.3) is written in the form:

$$\log f_2 = [\%\, 2]\frac{\partial \log f_2}{\partial\, [\%\, 2]} + [\%\, 3]\frac{\partial \log f_2}{\partial\, [\%\, 3]} + \ldots + [\%\, n]\frac{\partial \log f_2}{\partial\, [\%\, n]}. \qquad (9.1.7)$$

The partial derivatives are now represented by the symbols $e_2^{(2)}$, $e_2^{(3)}$, ..., $e_2^{(n)}$:

$$\log f_2 = [\%\, 2]\, e_2^{(2)} + [\%\, 3]\, e_2^{(3)} + \ldots + [\%\, n]\, e_2^{(n)}, \qquad (9.1.8)$$

while equation (9.1.6) must be replaced by

$$e_i^{(j)} = \frac{M_i}{M_j}\, e_j^{(i)}, \qquad (9.1.9)$$

where M_i and M_j are the atomic weights of i and j. Like the ε parameters the e parameters are constants for small concentrations of the solutes, i.e. concentrations for which Henry's law is obeyed.

9.2. Solubility of hydrogen, nitrogen and oxygen in liquid iron alloys

HYDROGEN

The solubility of hydrogen in a number of liquid binary iron alloys has been measured by Weinstein and Elliott [1]. For these solutions equation (9.1.8) takes the form:

$$\log f_{\mathrm{H}} = e_{\mathrm{H}}^{(\mathrm{H})}\, [\%\, \mathrm{H}] + e_{\mathrm{H}}^{(j)}\, [\%\, j], \qquad (9.2.1)$$

where j is the alloying element in iron and where f_{H} becomes equal to unity in the infinitely dilute solution.

Weinstein and Elliott found that the $\sqrt{p}$ law is obeyed by hydrogen in pure iron at all concentrations studied. This means that f_{H} remains one in all these Fe-H solutions and that $e_{\mathrm{H}}^{(\mathrm{H})}$ is zero. As a consequence, equation (9.2.1) reduces to:

$$\log f_{\mathrm{H}}^{(j)} = e_{\mathrm{H}}^{(j)}\, [\%\, j]. \qquad (9.2.2)$$

[1] M. WEINSTEIN and J. F. ELLIOTT, Trans. AIME **227**, 382 (1963). This article also contains a discussion of previous work on the solubility of hydrogen in liquid iron and liquid iron alloys.

The activity coefficient of hydrogen in any of the alloys was obtained by measuring the solubilities in pure iron and in the alloy at the same values of the hydrogen pressure p and the temperature T. The activity of hydrogen then has the same value in both solutions, so that, since $f_H^{(H)} = 1$:

$$a_H = [\% \text{ H (pure iron)}] = f_H^{(j)} [\% \text{ H (alloy)}]$$

or:

$$f_H^{(j)} = \left[\frac{\% \text{ H (pure iron)}}{\% \text{ H (alloy)}} \right]_{p,T} . \qquad (9.2.3)$$

Since Sieverts' law is found to be obeyed by hydrogen not only in pure iron but also in all iron alloys studied, $f_H^{(j)}$ is independent of the hydrogen pressure. It tends to unity as the alloy composition approaches that of pure iron.

Fig. 113 shows the effect of various alloying elements on the solubility of hydrogen in liquid iron at 1592 °C according to Weinstein and Elliott (l.c.). Carbon, boron, aluminium, germanium, tin, cobalt, copper, sulphur and phosphorus decrease the solubility while niobium, chromium and manganese increase it. Up to 16 wt% nickel has virtually no influence on

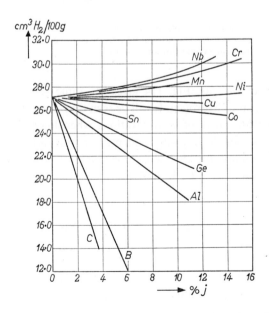

Fig. 113. Solubility of hydrogen at 1 atm H_2 pressure in liquid binary iron alloys at 1592 °C. The horizontal axis gives the concentration of the alloying element in weight percent, the vertical axis the number of cm³ (STP) hydrogen per 100 grams alloy (Weinstein and Elliott).

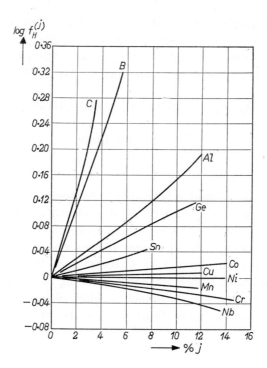

Fig. 114. Effect of several alloying elements on the activity coefficient of hydrogen in liquid binary iron alloys at 1592 °C. The horizontal axis gives the concentration of the alloying element in weight percent (Weinstein and Elliott).

the solubility of the gas. Fig. 114 gives the same results in terms of the activity coefficient of hydrogen. It will be seen that straight lines are obtained over a fairly wide range of concentrations when $\log f_{\mathrm{H}}$ is plotted against the concentration of the alloying element. For iron-carbon, iron-boron and iron-aluminium alloys (at 1592 °C), for example, the following relationships hold up to about 2 wt% C, B or Al:

$$\log f_{\mathrm{H}}^{(\mathrm{C})} = 0.060 \,[\% \,\mathrm{C}], \qquad (9.2.4)$$

$$\log f_{\mathrm{H}}^{(\mathrm{B})} = 0.050 \,[\% \,\mathrm{B}], \qquad (9.2.5)$$

$$\log f_{\mathrm{H}}^{(\mathrm{Al})} = 0.013 \,[\% \,\mathrm{Al}]. \qquad (9.2.6)$$

The numerical factor appearing in these relationships is, according to Section 9.1 and equation (9.2.2), none other than the parameter

$$e_{\mathrm{H}}^{(j)} = \frac{\partial \log f_{\mathrm{H}}}{\partial \,[\% \,j]} \quad \text{for} \quad [\% \,j] \to 0$$

where the symbol j indicates the alloying element. Table 17 gives the values of this parameter for a number of alloying elements according to Weinstein and Elliott (l.c.), and according to Maekawa and Nakagawa [1]).

TABLE 17

EFFECT OF ALLOYING ELEMENTS ON THE ACTIVITY COEFFICIENT OF HYDROGEN IN LIQUID IRON AT INFINITE DILUTION AND ABOUT 1600 °C.

Alloying element j	$e_H^{(j)}$ Weinstein and Elliott	$e_H^{(j)}$ Maekawa and Nakagawa
Al	+0.013	−0.006
B	+0.050	
C	+0.060	+0.05
Co	+0.0018	+0.005
Cr	−0.0022	−0.031
Cu	+0.0005	
Ge	+0.010	
Mn	−0.0014	+0.047
Nb	−0.0023	
Ni	0.000	−0.019
P	+0.011	
S	+0.008	−0.011
Sn	+0.0053	

The values of $e_H^{(j)}$ can be used to calculate those of $\varepsilon_H^{(j)}$ (see previous section) by means of the simple relationship

$$\varepsilon_H^{(j)} = \frac{M_j}{0.2425} e_H^{(j)},$$
(9.2.7)

where M_j represents the atomic weight of the alloying element j.

The effect of temperature on the solubility of hydrogen in the various alloys was measured by Weinstein and Elliott (l.c.) in the range of 1530° to 1750 °C. The heats of solution derived from these measurements fall in the range of 8 to 9 kcal per 0.5 mole H_2 for all the alloys studied with the exception of the iron-aluminium alloys.

Several investigators have discussed the possibility of a correlation between the effective free electron concentration of an alloying addition and its effect on the solubility of hydrogen in the base metal. This is based on the reason-

---

[1]) S. MAEKAWA and Y. NAKAGAWA, *The solubility of hydrogen in liquid iron and iron alloys*, Reports I and II of the Japanese Steel Works, March 1, 1961, quoted by Weinstein and Elliott (l.c.).

able supposition that the gas dissolves in the metal in the form of protons and electrons:

$$\tfrac{1}{2} H_2 \rightleftharpoons [H^+] + [e^-].$$

In the most simplified reasoning it is concluded that an increase in the electron concentration in the metal will lead to a reduction in the solubility of hydrogen and vice versa. In agreement with this, Weinstein and Elliott (l.c.) come to the conclusion on the basis of their experiments that alloying elements having a larger effective number of free electrons than iron decrease the solubility of hydrogen in the liquid metal, while elements having a smaller number of free electrons than iron increase it. There is little doubt, however, that the problem under discussion is too complicated to be discussed in this simplified manner.

NITROGEN

The solubility of nitrogen in a number of liquid binary iron alloys has been studied by Pehlke and Elliott [1]) and several other investigators. Nitrogen, like hydrogen, obeys the $\sqrt{p}$ law in these alloys (and in pure iron), so that relationships analogous to (9.2.2) and (9.2.3) are valid:

$$\log f_N^{(j)} = e_N^{(j)} [\% j], \tag{9.2.8}$$

$$f_N^{(j)} = \left[\frac{\% \text{ N (pure iron)}}{\% \text{ N (alloy)}} \right]_{p,T} . \tag{9.2.9}$$

Fig. 115 shows the effect of a number of alloying elements on the solubility of nitrogen in liquid iron at 1600 °C according to Pehlke and Elliott (l.c.). Fig. 116 gives the same results in terms of the activity coefficient of nitrogen. Table 18 gives the values of

$$e_N^{(j)} = \frac{\partial \log f_N}{\partial [\% j]} \quad \text{for} \quad [\% j] \to 0$$

for a number of alloying elements j according to Pehlke and Elliott (l.c.), Schenck et al. [2]), and Maekawa and Nakagawa [3]). The values of $\varepsilon_N^{(j)}$ (see

[1]) R. D. PEHLKE and J. F. ELLIOTT, Trans. AIME **218**, 1088 (1960). This article also contains many references to older literature on the solubility of nitrogen in liquid iron and liquid iron alloys.
[2]) H. SCHENCK, M. G. FROHBERG and H. GRAF, Arch. Eisenhüttenwes. **29**, 673 (1958); **30**, 533 (1959).
[3]) S. MAEKAWA and Y. NAKAGAWA, Tetsu to Hagane (J. Iron Steel Inst. Japan) **45**, 255 (1959), quoted by Pehlke and Elliott (l.c.).

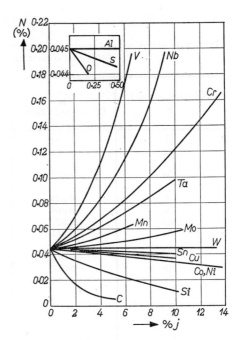

Fig. 115. Solubility of nitrogen at 1 atm
N₂ pressure in liquid binary iron alloys
at 1600 °C. The horizontal axis gives the
concentration of the alloying element in
weight percent, the vertical axis the
weight percentage of nitrogen (Pehlke
and Elliott).

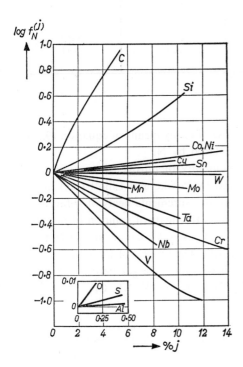

Fig. 116. Effect of several alloying elements on the activity coefficient of
nitrogen in liquid binary iron alloys at
1600 °C. The horizontal axis gives the
concentration of the alloying element in
weight percent (Pehlke and Elliott).

TABLE 18

EFFECT OF ALLOYING ELEMENTS ON THE ACTIVITY COEFFICIENT OF NITROGEN IN LIQUID
IRON AT INFINITE DILUTION AND 1600 °C.

Alloying element j	$e_N^{(j)}$ Pehlke and Elliott	$e_N^{(j)}$ Schenck et al.	$e_N^{(j)}$ Maekawa and Nakagawa
Al	+0.0025		+0.006
As		+0.018	
C	+0.25	+0.125	+0.135
Co	+0.011	+0.007	+0.005
Cr	−0.045		−0.057
Cu	+0.009	+0.002	
Mn	−0.02		−0.020
Mo	−0.011	−0.004	−0.013
Ni	+0.010	+0.010	+0.007
Nb	−0.067		
O	+0.050	0.0	−0.16
S		+0.013	
Sb		+0.009	
Se		0.000	
Si	+0.047	+0.065	+0.048
Sn	+0.007	+0.002	
Ta	−0.034		
Ti			−0.63
V	−0.10 *)		−0.11
W	−0.002		

Section 9.1) can be calculated from those of $e_N^{(j)}$ by means of equation (9.2.7) in which the subscript H has been replaced by N.

The effect of temperature on the solubility of nitrogen in a number of liquid binary iron alloys was measured by Pehlke and Elliott (l.c.) in the range of 1550° to 1775 °C. From these measurements heats of solution of nitrogen could be derived. Fig. 117 shows the heats of solution of nitrogen in the binary solutions containing V, Nb, Cr, Cu, W, Co or Ni, as functions of the concentration of the alloying element. The process of solution becomes more exothermal when the liquid iron contains a larger percentage of an alloying element with a great affinity for nitrogen (V, Nb, Cr).

---

*) Due to a misprint Table II of Pehlke and Elliott gives the incorrect value −0.010.

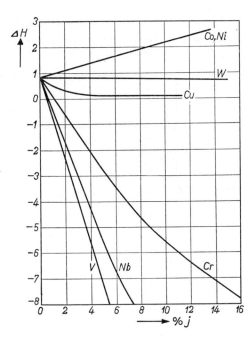

Fig. 117. Heat of solution (kcal per mole N_2) of nitrogen gas in liquid binary iron alloys as a function of the weight percentage of the alloying element (Pehlke and Elliott).

OXYGEN

The partial pressure of oxygen in equilibrium with a solution of oxygen in liquid iron or a liquid iron alloy is too low for direct measurement. As we have seen in preceding chapters, the activity of oxygen in the metal can be determined in this case by other means, e.g. by bringing the metal into equilibrium with a mixture of hydrogen and steam or with any slag having a known activity of oxygen.

In analogy with what was found for the solubility of nitrogen, the activity coefficient of oxygen in liquid iron is in general decreased (the solubility increased) by solutes which have a greater affinity for oxygen than iron (e.g. B, Al, C, Si, Ti, V, Nb, Cr), conversely it is generally increased by solutes with a smaller affinity for oxygen than iron (e.g. Mo, Co, Ni).

The values of

$$e_O^{(j)} = \frac{\delta \log f_O}{\delta \, [\% \, j]} \quad \text{for } [\% j] \to 0,$$

which can be found in the literature differ considerably from each other. Table 19 gives a survey of these values [1]. The value of $e_O^{(N)}$ has been cal-

---

[1] The table has partly been taken from a summary by C. BODSWORTH, *Physical Chemistry of Iron and Steel Manufacture*, Longmans, London 1963, p. 470.

TABLE 19

EFFECT OF ALLOYING ELEMENTS ON THE ACTIVITY COEFFICIENT OF OXYGEN IN LIQUID
IRON AT INFINITE DILUTION AND ABOUT 1600 °C.

Alloying element j	$e_O^{(j)}$
Al	-12.0 [1]), -1.0 [2])
Au	-0.005 [3])
B	-0.62 [4])
C	-0.485 [5]), -0.45 [6]), -0.41 [1]), -0.32 [7]), -0.13 [8])
Cr	-0.064 [9]), -0.041 [1])
Co	$+0.007$ [3])
Cu	-0.010 [3])
Mn	0 [1])
Mo	$+0.0035$ [3])
Ni	$+0.006$ [10])
N	$+0.057$ [11]), 0 [12])
Nb	-0.14 [13])
P	$+0.07$ [14]), -0.032 [15])
Pt	$+0.005$ [3])
Si	-0.16 [16]), -0.137 [17]), -0.087 [6]), -0.02 [1])
S	0 [18])
Ti	-0.187 [19])
W	$+0.009$ [3])

[1]) J. CHIPMAN, J. Iron Steel Inst. **180**, 97 (1955); N. A. GOKCEN and J. CHIPMAN, Trans. AIME **197**, 173 (1953)
[2]) J. C. D'ENTREMONT, D. L. GUERNSEY and J. CHIPMAN, Trans. AIME **227**, 14 (1963)
[3]) T. P. FLORIDIS and J. CHIPMAN, Trans. AIME **212**, 549 (1958)
[4]) C. YUAN HSI, Acta Met. Sinica **3**, 252 (1958)
[5]) E. T. TURKDOGAN, L. E. LEAKE and C. R. MASSON, Acta Met. **4**, 396 (1956)
[6]) H. SCHENCK and K. H. GERDOM, Arch. Eisenhüttenwes. **30**, 451 (1959)
[7]) S. BANYA and S. MATOBA, Tetsu to Hagane **44**, 643 (1958)
[8]) T. FUWA and J. CHIPMAN, Trans. AIME **215**, 708 (1959); **218**, 887 (1960)
[9]) E. T. TURKDOGAN, J. Iron Steel Inst. **178**, 278 (1954)
[10]) H. A. WRIEDT and J. CHIPMAN, Trans. AIME **206**, 1195 (1956)
[11]) R. D. PEHLKE and J. F. ELLIOTT, Trans. AIME **218**, 1088 (1960)
[12]) H. SCHENCK, M. G. FROHBERG and H. GRAF, Arch. Eisenhüttenwes. **29**, 673 (1958); **30**, 533 (1959)
[13]) M. ELLE and J. CHIPMAN, Trans. AIME **221**, 701 (1961)
[14]) D. DUTILLOY and J. CHIPMAN, Trans. AIME **218**, 428 (1960)
[15]) J. PEARSON and E. T. TURKDOGAN, J. Iron Steel Inst. **176**, 19 (1954)
[16]) J. CHIPMAN and T. C. M. PILLAY, Trans. AIME **221**, 1277 (1961)
[17]) S. MATOBA, K. GUNJI and T. KUWANA, Tetsu to Hagane **45**, 1328 (1959)
[18]) S. MATOBA and T. UNO, Tetsu to Hagane **28**, 651 (1942)
[19]) J. CHIPMAN, Trans. AIME **218**, 767 (1960)

culated from that of $e_N^{(O)}$ (Table 18) by means of equation (9.1.9), which in this case reads:

$$e_O^{(N)} = \frac{16}{14} e_N^{(O)}.$$

9.3. Chemical reaction between a dissolved gas and one of the components of a liquid alloy

In the last two sections it has been tacitly assumed that the concentrations of the dissolved gases and other solutes are such that no new phases can form in the alloys. The solutions considered of hydrogen in liquid iron-base alloys satisfy this condition under all circumstances: even under relatively high hydrogen pressures no hydride is formed in any of the alloys mentioned in Table 17. In the presence of oxygen or nitrogen, however, the possibility of the formation of an oxide or nitride must be taken into account. For example, let us consider a liquid iron-vanadium alloy brought into equilibrium with nitrogen of varying pressure.

Since nitrogen has a greater affinity for vanadium than for iron, its solubility in vanadium-bearing iron is greater than in pure iron. This has already been illustrated in Fig. 115 which shows the solubility of nitrogen at 1 atm as a function of the vanadium content at a temperature of 1600 °C. If we proceed from a particular Fe-V alloy and plot the nitrogen content as a function of the square root of the nitrogen pressure, we obtain a curve which

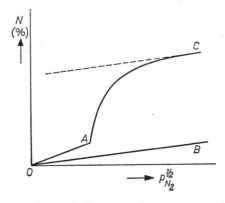

Fig. 118. Curve OAC shows schematically the nitrogen content of an Fe-X alloy as a function of the square root of the nitrogen pressure in the case where nitrogen has a greater affinity for X than for iron. At A the solubility limit has been reached. At higher nitrogen pressures a nitride precipitate forms in the alloy. The line OB gives the solubility of nitrogen in pure iron.

is shown schematically in Fig. 118. The first part of the curve (OA) is linear according to experimental data, i.e. the quantity of nitrogen absorbed obeys the $\sqrt{p}$ law. The increase in solubility due to the presence of vanadium is expressed in the steeper slope of the straight line compared to that for pure iron (OB).

At A the solubility limit of nitrogen in the alloy has been reached. If the nitrogen pressure increases still further a precipitate of vanadium nitride, VN, is formed. If the system consisted of only two components then, according to the phase rule, the pressure would remain constant from A until all the metal had been converted into the new phase. In the system in question, which consists of three components (Fe-V-N), the effective vanadium content of the solution is decreased as more and more vanadium reacts with nitrogen to form vanadium nitride. Therefore, after the solubility limit has been exceeded, the curve can be expected to take the form shown schematically by AC. This is such that the curve finally tends towards a straight line parallel to that for pure iron. The distance between the two parallel lines will be greater in proportion to the vanadium content of the alloy.

The influence of vanadium on the activity coefficient of nitrogen in iron follows from the solubility measurements below the solubility limit. Equation (9.2.9) shows that $f_N^{(V)}$ is given directly by the ratio of the slope of the pure iron line OB to the slope of the Fe-V line OA. El Tayeb and Parlee [1] have determined $f_N^{(V)}$ in this way for alloys with various vanadium contents and find at 1604 °C:

$$e_N^{(V)} = \frac{\delta \log f_N}{\delta\,[\%\,V]} = -0.094 , \qquad (9.3.1)$$

a value which deviates but little from the values given in Table 18.

At nitrogen pressures greater than that corresponding to point A (Fig. 118) we are dealing with a dissociation equilibrium which can be described by means of the formula

$$VN \rightleftharpoons [V] + \tfrac{1}{2}\,N_2, \qquad (9.3.2)$$

where the symbol [V] refers to vanadium dissolved in liquid iron. El Tayeb and Parlee (l.c.) studied this equilibrium for seven Fe-V alloys, of which the vanadium contents (in wt%) are given by the first column in Table 20. The second column gives the square root of the nitrogen pressure (in atm) at which a precipitate of VN begins to form at 1604 °C (point A in Fig. 118). The vanadium contents are much too large to be able to write the equilibrium

---

[1] N. M. EL TAYEB and N. A. D. PARLEE, Trans. AIME **227**, 929 1963).

constant of reaction (9.3.2) simply as the product of $[\% \text{ V}]$ and $\sqrt{p_{N_2}}$. Indeed, it is immediately obvious from the table that this product does not have a constant value. As we have already seen, in this case we must write

$$K = f_V [\% \text{ V}] \sqrt{p_{N_2}}, \qquad (9.3.3)$$

where f_V, according to equation (9.1.2), can be written approximately as

$$f_V = f_V^{(V)} f_V^{(N)}. \qquad (9.3.4)$$

Columns 3 and 4 in Table 20 give information about the values of these activity coefficients. Column 5 gives the product of the numbers in the first four columns, i.e. the value of the equilibrium constant at 1604 °C.

TABLE 20

DATA RELATING TO THE EQUILIBRIUM VN $\rightleftharpoons$ [V] + $\frac{1}{2}$ N$_2$ AT 1604 °C ACCORDING TO EL TAYEB AND PARLEE

% V in liquid iron	$\sqrt{p_{N_2}}$	$f_V^{(N)}$	$f_V^{(V)}$	K
8.0	0.815	0.84	1.49	8.2
9.0	0.696	0.83	1.55	8.1
10.0	0.620	0.83	1.61	8.3
10.8	0.554	0.82	1.65	8.1
11.5	0.518	0.82	1.70	8.3
12.6	0.442	0.81	1.77	8.0
15.0	0.279	0.81	—	—

The dissociation equilibrium in question can also be described by means of the equation

$$\text{VN} \rightleftharpoons [\text{V}] + [\text{N}], \qquad (9.3.5)$$

where the symbols [V] and [N] refer to the concentrations of vanadium and nitrogen in the saturated solution. At a vanadium content of 8% and at 1604 °C (Table 20) the solution is saturated at a nitrogen content of 0.214%. When this concentration is exceeded precipitation of VN occurs.

In some cases even minute quantities of a dissolved gas are sufficient to produce a new phase. An extreme case is the formation of Al$_2$O$_3$ in iron containing aluminium and oxygen according to the reaction

$$2[\text{Al}] + 3[\text{O}] \rightleftharpoons \text{Al}_2\text{O}_3. \qquad (9.3.6)$$

For iron with a small aluminium content one may write the equilibrium constant for the reverse reaction to a reasonable approximation as

$$K = [\% \text{ O}]^3 [\% \text{ Al}]^2. \qquad (9.3.7)$$

At 1600 °C constant K is of the order of magnitude of 10^{-14} if the symbol %
refers to weight percent [1]). If liquid iron contains 0.003 % Al it can therefore
contain at most 0.001 % O in solution. At higher oxygen contents the solubility product given by (9.3.7) is exceeded and a precipitate of Al_2O_3 forms in
the alloy. As the temperature rises the solubility product, and thus the
solubility of Al_2O_3, increases rapidly.

The above shows that the solubility of VN in liquid iron is large, while
that of Al_2O_3 is very small. The solubility of aluminium nitride, AlN, lies
in between. According to Wiester et al. [2]) the solubility product

$$K = [\% \text{ Al}] [\% \text{ N}]$$

at 1620 °C and small aluminium contents has the value 6.10^{-3}, if the symbol
% indicates weight percentages. Thus, if liquid iron contains 0.2 % Al, precipitation of AlN will occur as soon as the nitrogen content exceeds a value
of 0.03 %.

9.4. Solubility and precipitation of some mononitrides in solid iron

In solid iron containing vanadium or aluminium, much smaller quantities
of nitrogen are required to produce a precipitate of VN or AlN than would
be needed in the liquid alloys. In other words: the solubilities of VN and
AlN are much smaller in solid iron than in liquid iron. The same is true for
Al_2O_3; its solubility in solid iron is practically zero. In this section we shall
discuss not only the solubility of VN and AlN, but also that of BN and NbN
in solid iron.

VANADIUM NITRIDE

The schematic Figure 118 does not only apply to liquid alloys Fe-X-N in
which an X nitride can precipitate, but also to the corresponding solid alloys.
This is demonstrated by Fig. 119 which shows the results of measurements
by Fountain and Chipman [3]) on the relationship between the nitrogen con-

---

[1]) J. C. D'ENTREMONT, D. L. GUERNSEY and J. CHIPMAN, Trans. AIME **227**, 14 (1963)
[2]) H. J. WIESTER, W. BADING, H. RIEDEL and W. SCHOLZ, Stahl Eisen **77**, 773 (1957)
[3]) R. W. FOUNTAIN and J. CHIPMAN, Trans. AIME **212**, 737 (1958)

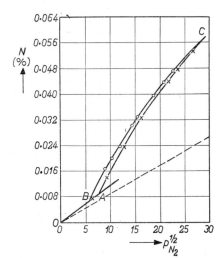

Fig. 119. Nitrogen content of fcc iron containing 0.17 wt% vanadium as a function of the square root of the nitrogen pressure (mm of mercury) at a temperature of 1050 °C. The broken line is given for comparison and relates to pure iron (Fountain and Chipman).

tent (wt%) and square root of the equilibrium nitrogen pressure for fcc iron containing 0.17% vanadium and a temperature of 1050 °C. As the pressure increases curve OAC is described, with decreasing pressure curve CBO. The authors were able to show that this hysteresis is a particle size effect. When point A is reached VN particles of very small size begin to precipitate and are in metastable equilibrium with a solid solution of higher nitrogen content than that corresponding to the stable equilibrium with big particles. As the pressure of nitrogen is increased the particles grow along the curve AC. On removing nitrogen from the system, decomposition first of the smaller and finally of the larger particles takes place along the curve CB, resulting in a lower solubility. Consequently, the true equilibrium solubility is not given by point A but is approached more nearly by point B.

As in the case of the liquid alloys (preceding section) the effect of vanadium on the activity coefficient of nitrogen is obtained from the solubility determinations up to the beginning of precipitation. As would be expected, Fig. 120 shows that the effect of vanadium is greater when the temperature is lower. The straight lines correspond to the equations:

$$\log f_N^{(V)} = -2.2 \ [\% \, V] \qquad (750 \, °C),$$

$$\log f_N^{(V)} = -0.47 \ [\% \, V] \qquad (950 \, °C),$$

$$\log f_N^{(V)} = -0.33 \ [\% \, V] \qquad (1050 \, °C),$$

$$\log f_N^{(V)} = -0.18 \ [\% \, V] \qquad (1200 \, °C).$$

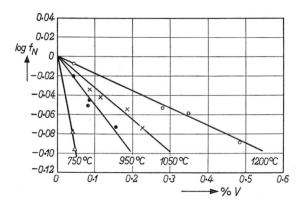

Fig. 120. Influence of vanadium on the activity coefficient of nitrogen in solid iron (Fountain and Chipman).

In order to determine the equilibrium constant of the reaction

$$\text{VN} \rightleftharpoons [\text{V}] + \tfrac{1}{2} \text{N}_2 \qquad (9.4.1)$$

Fountain and Chipman determined the position of point B (Fig. 119) for a number of alloys with different vanadium contents and for different temperatures. In all the cases investigated B corresponded to a vanadium content lying between 0.04 and 0.5% and a nitrogen content lower than 0.01%. These contents are so low that both $f_\text{V}^{(\text{V})}$ and $f_\text{V}^{(\text{N})}$ can be equated to unity. That this is justified can be shown directly for $f_\text{V}^{(\text{N})}$ with the help of the reciprocal relation (9.1.9) and the data in Fig. 120. According to equation (9.3.4) the activity coefficient of vanadium can be equated to unity. The equilibrium constant of (9.4.1) for the solid alloys under discussion can thus simply be written

$$K = [\% \text{ V}] \sqrt{p_{\text{N}_2}}. \qquad (9.4.2)$$

If p is expressed in mm of mercury and the vanadium content in weight percent, then K as a function of the absolute temperature for γ and α iron has the following values:

$$\log K(\gamma) = -\frac{7500}{T} + 5.65 , \qquad (9.4.3)$$

$$\log K(\alpha) = -\frac{6250}{T} + 4.77 . \qquad (9.4.4)$$

From the experimental data of Fountain and Chipman (l.c.) follow also values for the equilibrium constant of the reaction

$$VN \rightleftharpoons [V] + [N],\qquad(9.4.5)$$

i.e. values for the solubility product

$$K_s = [\% \, V]f_N\,[\% \, N].\qquad(9.4.6)$$

Here $[\% \, V]$ and $[\% \, N]$ are the concentrations in wt% in the saturated solution, while f_N is the activity coefficient of nitrogen, given by Fig. 120. As a function of the absolute temperature K_s for γ and α iron has the following values:

$$\log K_s(\gamma) = -\frac{7070}{T} + 2.27 \,,\qquad(9.4.7)$$

$$\log K_s(\alpha) = -\frac{7830}{T} + 2.45 \,.\qquad(9.4.8)$$

ALUMINIUM NITRIDE

According to Section 9.3 the solubility of AlN in *liquid* iron is smaller than that of VN. The same is true for the solubilities of these compounds in *solid* iron.

Darken, Smith and Filer [1]) determined the equilibrium constant of the reaction

$$AlN \rightleftharpoons [Al] + [N]\qquad(9.4.9)$$

for solutions of aluminium and nitrogen in γ iron. The total aluminium content in all experiments was less than 0.15 wt%. As an approximation they write the equilibrium constant (the solubility product) in the form

$$K_s = [\% \, Al]\,[\% \, N].\qquad(9.4.10)$$

From measurements at 1050°, 1200°, and 1350 °C they find the following dependence of K_s on the absolute temperature:

$$\log K_s(\gamma) = -\frac{7400}{T} + 1.95.\qquad(9.4.11)$$

Other investigators find that the dependence of K_s on the temperature differs somewhat from (9.4.11). According to Shimose and Narita [2]) this depend-

[1]) L. S. DARKEN, R. P. SMITH and E. W. FILER, Trans. AIME **191**, 1174 (1951)
[2]) T. SHIMOSE and K. NARITA, Tetsu to Hagane **40**, 242 (1954)

ence is given by

$$\log K_s(\gamma) = -\frac{7184}{T} + 1.786 \,, \qquad (9.4.12)$$

according to König, Scholz und Ulmer [1]) by

$$\log K_s(\gamma) = -\frac{7750}{T} + 1.8 \,. \qquad (9.4.13)$$

Heats of reaction can be derived without difficulty from equations (9.4.3), (9.4.4), (9.4.7), (9.4.8), (9.4.11), (9.4.12) and (9.4.13).

BORON NITRIDE

The solubility of boron in solid iron is extremely small. If an Fe-B alloy contains more boron than corresponds to this solubility, then the compound Fe_2B is present as a second solid phase. With gradually rising nitrogen pressure an iron-boron alloy containing Fe_2B, like a single-phase alloy, first absorbs nitrogen in solid solution in the iron. When a certain nitrogen content is exceeded the nitride BN begins to form, so that one is dealing with a four-phase equilibrium between iron (containing B and N in solution), BN, Fe_2B and gas. According to the phase rule this system has only one degree of freedom. At constant temperature and rising nitrogen content the nitrogen pressure will therefore remain constant until one of the phases, in this case Fe_2B, has disappeared.

The foregoing is clearly demonstrated by Fig. 121, which relates to measurements at 1050 °C by Fountain and Chipman [2]) on an Fe-B alloy containing 0.013 wt% B, while the solubility at 1050 °C in γ iron is only 0.006% [3]). At point A the formation of BN commences, after which the nitrogen content of the alloy continues to increase at constant pressure. After all the Fe_2B has disappeared, at point B, the nitrogen content increases still further at rising pressure. As soon as the maximum amount of boron is removed from solid solution, the curve becomes nearly parallel to the broken line relating to pure iron.

Point A, at which the first break in the curve occurs, establishes the equilibrium constant for the four-phase reaction

[1]) P. König, W. Scholz and H. Ulmer, Arch. Eisenhüttenwes. **32**, 541 (1961)
[2]) R. W. Fountain and J. Chipman, Trans. AIME **224**, 599 (1962)
[3]) M. E. Nicholson, Trans. AIME **200**, 185 (1954)

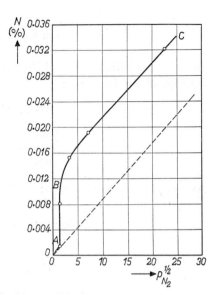

Fig. 121. Relationship between the nitrogen content and square root of the equilibrium nitrogen pressure (in mm Hg) at 1050 °C for an Fe-B alloy containing 0.013 %B. The broken line relates to pure iron (Fountain and Chipman).

$$2 \, \text{Fe}(\gamma) + \text{BN} \rightleftharpoons \text{Fe}_2\text{B} + \tfrac{1}{2} \, \text{N}_2, \qquad (9.4.14)$$

$$K_{14} = \sqrt{p_{\text{N}_2}}. \qquad (9.4.15)$$

If p is expressed in atm, then the experimental results of Fountain and Chipman (l.c.) are given by the equation

$$\log K_{14} = -\frac{5700}{T} + 2.89 \, . \qquad (9.4.16)$$

Since the quantity of boron which is present in solution at point A is known, the experimental results also yield directly the value of the equilibrium constant K_{17} for the reaction

$$\text{BN} \rightleftharpoons [\text{B}]_\gamma + \tfrac{1}{2} \, \text{N}_2, \qquad (9.4.17)$$

$$K_{17} = [\% \, \text{B}]_\gamma \sqrt{p_{\text{N}_2}} \, , \qquad (9.4.18)$$

$$\log K_{17} = -\frac{14400}{T} + 7.20 \, . \qquad (9.4.19)$$

The solubility product of BN in γ iron can be derived from the above by also employing the equilibrium constant of the reaction

$$\tfrac{1}{2} \, \text{N}_2 \rightleftharpoons [\text{N}]_\gamma \, , \qquad (9.4.20)$$

which according to equation (7.8.7) is given by

$$\log K_{20} = \frac{450}{T} - 1.96 , \qquad (9.4.21)$$

By adding (9.4.17) and (9.4.20) we get equation (9.4.22) describing the solution of BN in γ iron:

$$BN \rightleftharpoons [B]_\gamma + \tfrac{1}{2} N_2 \qquad (9.4.17)$$

$$\tfrac{1}{2} N_2 \rightleftharpoons [N]_\gamma \qquad (9.4.20)$$

$$\overline{BN \rightleftharpoons [B]_\gamma + [N]_\gamma} \qquad (9.4.22)$$

As a consequence, we obtain:

$$\log K_{17} = -\frac{14400}{T} + 7.20 \qquad (9.4.19)$$

$$\log K_{20} = \frac{450}{T} - 1.96 \qquad (9.4.21)$$

$$\overline{\log K_{22} = -\frac{13950}{T} + 5.24 ,} \qquad (9.4.23)$$

where

$$K_{22} = [\% B]_\gamma [\% N]_\gamma. \qquad (9.4.24)$$

Finally, the amount of nitrogen in solid solution in γ iron in equilibrium with BN and Fe_2B follows from

$$2 Fe(\gamma) + BN \rightleftharpoons Fe_2B + \tfrac{1}{2} N_2 \qquad (9.4.14)$$

$$\tfrac{1}{2} N_2 \rightleftharpoons [N]_\gamma \qquad (9.4.20)$$

$$\overline{2 Fe(\gamma) + BN \rightleftharpoons Fe_2B + [N]_\gamma} \qquad (9.4.25)$$

Thus we obtain:

$$\log K_{14} = -\frac{5700}{T} + 2.89 \qquad (9.4.16)$$

$$\log K_{20} = \frac{450}{T} - 1.96 \qquad (9.4.21)$$

$$\overline{\log K_{25} = -\frac{5250}{T} + 0.93 ,} \qquad (9.4.26)$$

where $K_{25} = [\% N]_\gamma$ in equilibrium with BN and Fe_2B. From the various equations of the type (9.4.26) one can again directly derive heats of reaction.

NIOBIUM NITRIDE

The solubility of niobium nitride in γ iron (for the temperature range 1191° to 1336 °C)was determined by R. P. Smith [1]) by the same method as that used by Darken, Smith and Filer (l.c.) to determine the solubility of aluminium nitride in γ iron. The niobium contents of the Fe-Nb alloys investigated all lay below 1 wt%. The composition of the nitride which formed after the solubility product was exceeded was between NbN and $NbN_{0.9}$ (cf. Section 8.6).

As an approximation Smith writes the solubility product of niobium nitride in γ iron in the form

$$K_s = [\% \, Nb] \, [\% \, N]. \qquad (9.4.27)$$

It is given as a function of the temperature by

$$\log K_s = - \frac{10230}{T} + 4.04 , \qquad (9.4.28)$$

corresponding to an enthalpy (heat) of solution of 46800 cal/mole for the dilute solutions of NbN in the γ phase of solid iron.

9.5. Nitrogen in silicon iron

INTRODUCTION

Nitrogen can play an important part in silicon steel that is used as the "iron" core of transformers, electric motors and electric generators. The steel which is used in transformers generally contains about 3% silicon.

In the earliest transformers the core was made of unalloyed steel which was soft-annealed to make it as far as possible free of internal stresses. The coercivity and hysteresis losses in this material were relatively large as a result of the many inclusions it contained, especially inclusions of iron carbide. Also the eddy losses were relatively high due to the small electrical resistivity. It was especially unfortunate that the coercivity and hysteresis losses spontaneously increased with the course of time. This "magnetic ageing" was caused by the slow precipitation of nitrogen in the form of iron nitride (see Section 7.13).

---

[1]) R. P. SMITH, Trans. AIME **224**, 190 (1962)

The silicon steel used in transformers suffers smaller eddy losses than unalloyed steel due to its greater resistivity. Also the hysteresis losses are smaller because silicon encourages the formation of graphite which is magnetically less harmful than iron carbide, since for the same number of carbon atoms the total volume of inclusions is much smaller. It is also of importance that silicon steel exhibits no ageing phenomena because the nitrogen which is present as an impurity occurs, after a suitable annealing treatment, in the form of a very stable silicon nitride [1]). The composition of this nitride is Si_3N_4 [2,3]). However, this precipitate too must be considered undesirable since, like all other precipitates, it has an unfavourable effect on the coercivity and the hysteresis losses. In the next section, however, we shall see that by the deliberate addition of nitrogen to silicon steel one can profit from the presence of this element to obtain a magnetically favourable crystal orientation.

SOLUBILITY OF NITROGEN IN SILICON IRON IN EQUILIBRIUM WITH Si_3N_4

Si_3N_4 occurs in two crystal modifications which are indicated by $\alpha\ Si_3N_4$ and $\beta\ Si_3N_4$. Their structures have been determined by various investigators [4-7]). According to Seybolt [8]), in nitrided silicon steels at 900 °C only $\alpha\ Si_3N_4$ is present (investigated with a high-temperature X-ray diffraction apparatus). His experiments indicate that this is valid not only for 900 °C but also for the whole temperature range from 700° to 950 °C. However, when the alloys are cooled to room temperature and the nitrides extracted, both α and $\beta\ Si_3N_4$ are found.

For the solubility of nitrogen in silicon iron in equilibrium with Si_3N_4 various investigators give strongly divergent values. From the best experimental determinations carried out before 1959 [9-11]) Leslie [12]) derives the following equation for this solubility

[1]) J. D. Fast, Philips techn. Rev. **16**, 341 (1955)
[2]) W. C. Leslie, K. G. Carroll and R. M. Fisher, Trans. AIME **194**, 204 (1952)
[3]) H. A. Sloman, J. Iron Steel Inst. **182**, 307 (1956)
[4]) D. Hardie and K. H. Jack, Nature **180**, 332 (1957)
[5]) O. Glemser, K. Beltz and P. Naumann, Z. anorg. allg. Chem. **291**, 51 (1957)
[6]) S. N. Ruddlesden and P. Popper, Acta Cryst. **11**, 465 (1958)
[7]) K. Narita and K. Mori, Bull. Chem. Soc. Japan, **32**, 417 (1959)
[8]) A. U. Seybolt, Trans. AIME **212**, 161 (1958)
[9]) R. Rawlings, J. Iron Steel Inst. **184**, 53 (1956). See also Rawlings' later work: R. Rawlings and P. M. Robinson, J. Iron Steel Inst. **197**, 306 (1961)
[10]) N. S. Corney and E. T. Turkdogan, J. Iron Steel Inst. **180**, 344 (1955); E. T. Turkdogan and S. Ignatowicz, J. Iron Steel Inst. **185**, 200 (1957)
[11]) H. F. Beeghly, Anal. Chem. **24**, 1075 (1952)
[12]) W. C. Leslie, *Nitrogen in ferritic Steels*, New York (1959), Amer. Iron Steel Inst.

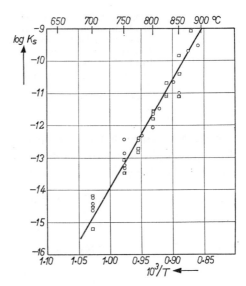

Fig. 122. Solubility of nitrogen in silicon-bearing α iron in equilibrium with Si_3N_4. The figure gives log K_s where $K_s = [\% \, Si]^3 [\% \, N]^4$ as a function of the reciprocal value of absolute temperature. The circles relate to Fe-Si alloys containing 0.48 % Si, the squares to alloys containing 0.75 % Si. (Fast and Verrijp, previously unpublished).

$$\log K_s = -\frac{20800}{T} + 9.9 , \qquad (9.5.1)$$

where

$$K_s = [\text{wt} \% \, Si]^3 [\text{wt} \% \, N]^4. \qquad (9.5.2)$$

Our own (previously unpublished) experiments [1]) lead to the equation

$$\log K_s = -\frac{33300}{T} + 19.4 , \qquad (9.5.3)$$

corresponding to a heat of solution of Si_3N_4 in α iron of 152 kcal/mole (see Fig. 122).

Our experiments were carried out with Fe-Si alloys containing 0.48 and 0.75 % Si. They were prepared by vacuum melting [2]) from pure iron and transistor-quality silicon. A quantity of about 0.02 % nitrogen was introduced into a great number of wires of this material by heating them at 950 °C in pure nitrogen to which 2 % hydrogen had been added. Then, by protracted heat treatment at 600 °C in pure, hydrogen-free nitrogen (in which there is no nitrogen interchange between metal and gas phase; cf. Section 7.8) all the nitrogen absorbed at 950 °C was converted into Si_3N_4. The concentration of dissolved nitrogen in equilibrium with Si_3N_4, which is negligibly small at 600 °C, was determined with rising

[1]) J. D. Fast and M. B. Verrijp, previously unpublished.
[2]) J. D. Fast, A. I. Luteijn and E. Overbosch, Philips techn. Rev. **15**, 114 (1953)

temperature by measurements of internal friction. At all temperatures at which an appreciable Snoek damping occurred, the damping peak was of the simple Debye type, i.e. could be specified by means of only one relaxation time. The maximum lay at the same temperature as in silicon-free Fe-N alloys. These facts were also ascertained for Fe-Si-N alloys with higher Si contents. Contrary to these findings, other investigators [1,2]) have noted two or more damping peaks. We suppose that the extra peaks must be ascribed to the presence of one or more impurities (carbon?) in their Fe-Si-N alloys.

We can visualize that a solution of Si_3N_4 in ferrite is formed by first decomposing the nitride into solid silicon and gaseous nitrogen and sub-sequently dissolving these elements in the metal:

$$Si_3N_4 \rightarrow 3\ Si + 2\ N_2 \qquad\qquad (9.5.4)$$

$$3\ Si \rightarrow 3\ [Si] \qquad\qquad (9.5.5)$$

$$2\ N_2 \rightarrow 4\ [N] \qquad\qquad (9.5.6)$$

$$Si_3N_4 \rightarrow 3\ [Si] + 4\ [N] \qquad\qquad (9.5.7)$$

According to Pehlke and Elliott [3]), when the first reaction (dissociation of α Si_3N_4) proceeds isothermally, 173 kcal/mole of heat is absorbed. The fact that reaction (9.5.7), according to the above, requires less heat (152 kcal), despite the fact that reaction (9.5.6) also is endothermal, depends on the strongly exothermal character of reaction (9.5.5). The thermal effect accompanying this reaction can be deduced from the work of Seybolt [4]).

Seybolt studied the equilibrium

$$Si_3N_4 \rightleftharpoons 3\ [Si] + 2\ N_2 \qquad\qquad (9.5.8)$$

by direct measurement of nitrogen pressures at various temperatures. He did this for Fe-Si-N alloys with various Si contents and with nitrogen contents, nearly all of which lay below 0.1 %. For a Si content of 3.3 % the dissociation pressure of reaction (9.5.8) reaches a value of 1 atm at 944 °C [5]). When $\log p(N_2)$ is plotted against $1/T$ a straight line is obtained with a slope of -9740 °K. The heat of reaction (9.5.8) is thus $2 \times 4.575 \times 9.74 = 89$ kcal, i.e. 84 kcal less than that of reaction (9.5.4). Therefore, when 1 gramatom

[1]) D. A. LEAK, W. R. THOMAS and G. M. LEAK, Acta Met. 3, 501 (1955)
[2]) R. RAWLINGS and P. M. ROBINSON, J. Iron Steel Inst. 197, 211 and 306 (1961)
[3]) R. D. PEHLKE and J. F. ELLIOTT, Trans. AIME 215, 781 (1959)
[4]) A. U. SEYBOLT, Trans. AIME 212, 161 (1958)
[5]) A lower temperature (770 °C) is found by M. L. PEARCE, Trans. AIME 227, 1393 (1963), who experimented with 3.06 % Si-Fe alloys.

of silicon is dissolved in 3.3 % silicon iron, 84/3 = 28 kcal heat is developed. In other words: the partial molar enthalpy of silicon in 3.3 % silicon iron relative to pure solid silicon as standard state of reference is —28 kcal per gr.-atom silicon.

From the heats of reactions (9.5.4), (9.5.5) and (9.5.7), which were seen above to amount to 173, —84 and 152 kcal, one can calculate a value of 63 kcal for reaction (9.5.6), i.e. a heat of solution of nitrogen in silicon iron of 16 kcal per gr.-atom nitrogen. This value is not very reliable considering the spread of the points in Fig. 122 and the fact that Seybolt experimented with alloys with Si contents different from ours. However, it does seem justifiable to conclude qualitatively that the solution of nitrogen in silicon iron is accompanied by a greater heat absorption than its solution in pure alpha iron which, according to equation (7.8.4), is 7.2 kcal per gr.-atom nitrogen. In accordance with this we shall see in the following that the solubility of nitrogen in silicon iron in equilibrium with N_2 is smaller than that in pure iron.

> With regard to the above it should be noted that according to Seybolt (l.c.) the nitrogen dissociation pressures in question correspond to metastable equilibria (9.5.8), insofar as they relate to Fe-Si-N alloys with Si-contents between about 1.3 and 4.4 %. He comes to this conclusion on the basis of his finding that the equilibrium pressures are much higher after cold-working or hot-working the alloys. He interprets this anomaly as indicating the occurrence of an order-disorder transformation, even in alloys containing as little as 1.5 % silicon.

SOLUBILITY OF NITROGEN IN SILICON IRON IN EQUILIBRIUM WITH A GAS PHASE OF KNOWN NITROGEN ACTIVITY

The solubility of gaseous nitrogen in silicon iron can be studied at temperatures which are so high and N_2 pressures which are so low that, for *thermodynamic* reasons, no Si_3N_4 can form. In place of gaseous nitrogen one can, of course, also use mixtures of gases with known nitrogen activities. Working by this method, several investigators [1,2] found that silicon decreases the solubility of nitrogen in solid iron. This agrees with the earlier discussed effect of silicon on the solubility of nitrogen in molten iron (see Fig. 115).

We have found that it is also possible to study the solubility of nitrogen in silicon iron at temperatures which are so low that, for *kinetic* reasons, no

[1] N. S. CORNEY and E. T. TURKDOGAN, J. Iron Steel Inst. **180**, 344 (1955)
[2] R. RAWLINGS, J. Iron Steel Inst. **185**, 441 (1957)

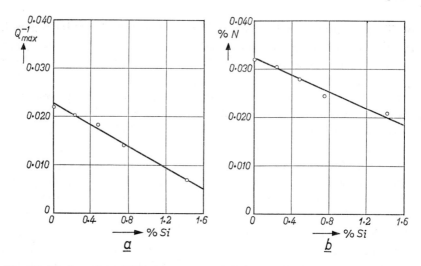

Fig. 123. Maximum internal friction (*a*) and analysed nitrogen percentage (*b*) as functions of the Si content of Fe-Si alloys after establishment of metastable equilibrium at 500 °C between the metal and a gas mixture at 1 atm consisting of 86% H_2 + 14% NH_3. At constant silicon content the maximum internal friction is proportional to the percentage of dissolved nitrogen (Fast and Verrijp, previously unpublished).

Si_3N_4 forms. These temperatures must, however, be high enough for nitrogen (in contrast to silicon) to have a relatively high diffusion rate in the metal. We chose for our experiments [1]) recrystallized wires with Si contents of 0.00, 0.24, 0.48, 0.75 and 1.42%, which were heated at a temperature of 500 °C in a current of gas composed of 86% H_2 + 14% NH_3. The heat treatment was continued until no further change occurred in the internal friction of the wires and in the result of the chemical analysis (Kjeldahl) of the quantity of nitrogen absorbed. Fig. 123 gives the results of these measurements.

The fact that the solubility of nitrogen in silicon iron decreases with increasing Si content, despite the fact that silicon has a much greater affinity for nitrogen than iron, shows that the "steric" factor is more important here than the purely "chemical" factor. The lattice constant of iron decreases when silicon is introduced and the accommodation of a nitrogen atom in an interstice of silicon iron therefore costs more elastic deformation energy than its accommodation in an interstice of pure iron.

The solutions in question of nitrogen in silicon iron are metastable. In the equilibrium state at 500 °C all the nitrogen present would be in the form of

---

[1]) J. D. Fast and M. B. Verrijp, not previously published.

Si_3N_4. In reality, as the temperature drops, it is not this most stable state which is established, but as a result of the comparative immobility of the Si atoms precipitation occurs as in the binary Fe-N solutions: Fe_4N forms above about 300 °C, while at even lower temperatures Fe_8N forms as intermediate and Fe_4N as final precipitate (cf. Section 7.10). These observations are in agreement with those of other investigators [1,2]).

9.6. Influence of a finely divided nitride in metals and alloys

INTRODUCTION

An important development in the application of metals is the deliberate employment of a finely divided second phase. This may be done with various aims in mind: (a) hardening of metals and alloys, (b) increasing their creep resistance, (c) prevention of crystal growth, (d) obtaining the desired orientation or form of crystals. To attain one of more of these aims, use is made preferably of a phase which dissolves in the solid metal at high temperatures and precipitates at lower temperatures. By means of rapid cooling from a high temperature it is then possible to obtain a supersaturated solution and the degree of dispersion can be controlled by the choice of temperature and duration of precipitation. Not only the size of the precipitated particles but also their shape and their distribution can be of importance.

We shall discuss in this section a few useful applications of nitride precipitates and in particular the grain-refining action of aluminium nitride in unalloyed steels and the role of silicon nitride and other compounds in obtaining a favourable crystal texture in silicon iron.

THE ROLE OF ALUMINIUM NITRIDE IN STEELS

In Section 7.13 the hardening of a metal by a finely divided precipitate was discussed, using as example the precipitation of iron nitride in nitrogen-bearing iron or mild steel. The resulting changes in the properties of mild steel are usually undesirable. Even less desirable are the phenomena of strain

[1]) J. JEZEK, J. VOBORIL and V. CIHAL, J. Iron Steel Inst. **195**, 49 (1960)
[2]) D. J. F. NANCARROW and R. RAWLINGS, Metal Treatment **29**, 175 (1962)

ageing caused by nitrogen, phenomena which will not be discussed until the second volume of this work [1]). In many cases aluminium (less than 0.1 %) is added to the steel to bind the nitrogen. Not only does this prevent the harmful action of nitrogen, but the addition of aluminium has furthermore the very advantageous effect that the grain growth at high temperatures is slowed down to a considerable degree. This is of great importance since the mechanical properties of steels are largely determined by the microstructure.

Many contradictory hypotheses have been put forward to explain the above-mentioned favourable effect of aluminium in mild steel (and also in medium-carbon steel). They involved the action of aluminium oxide, aluminium sulphide or segregation of aluminium in solution at austenite grain boundaries. From the experiments of many investigators, of which we shall mention only a few [2-5]), it finally became clear that aluminium nitride is the effective agent. As an illustration of this sort of investigation we shall discuss very briefly the work of Pomey and Vigneron (l.c.). They proceeded from a number of steels of average composition 0.41 % C, 0.45 % Si, 0.62 % Mn, 0.032 % S, 0.018 % P, balance Fe. These were melted and cast in a good vacuum. During the preparation 0.08 % Al was added to one of them, 0.03 % N (in the form of Fe_4N) to another and 0.08 % Al + 0.03 % N to a third. To a fourth Al + FeO was added. The grain growth at high temperatures was only inhibited to a large degree in the steel to which aluminium plus nitrogen had been added. Aluminium alone, nitrogen alone and aluminium oxide were found to have little or no influence.

The fact that AlN and not Al_2O_3 is the active constituent which discourages grain growth, is due, in our opinion, to the stability of Al_2O_3 being so great that it forms already in the melt (see Section 9.3). It can therefore coagulate and becomes too coarse to be effective. On the other hand AlN begins to form during cooling only after solidification of the metal. A grain refining compound in steel must satisfy the condition that it is not so stable that it already forms in the liquid, but, on the other hand, is stable enough that it only redissolves in the solid metal at high temperatures. Before redissolving, coalescence of the precipitate occurs, which makes it less effective. Al_2O_3 in a very finely divided state, because of its insolubility in solid steel, would be

[1]) J. D. FAST, *Interaction of Metals and Gases*, II. *Kinetics and Mechanisms*, to be published
[2]) E. HOUDREMONT and H. SCHRADER, Arch. Eisenhüttenwes. **12**, 393 (1938/39)
[3]) K. BORN and W. KOCH, Stahl Eisen **72**, 1268 (1952)
[4]) W. C. LESLIE, R. L. RICKETT, C. L. DOTSON and C. S. WALTON, Trans. Amer. Soc. Met. **46**, 1470 (1954)
[5]) J. POMEY and G. VIGNERON, Comptes rendus **249**, 1661 (1959)

preferable to AlN, but the necessary state of dispersion can be accomplished only by powder metallurgy or internal oxidation (see Section 9.7).

In the equilibrium state steel contains precipitated AlN as long as the temperature is low enough for the observed product [% Al] [% N] to be larger than the corresponding solubility product. For the control of grain size aluminium has the advantage over vanadium that the solubility of AlN in iron according to Section 9.4 is smaller than that of VN. For dilute solutions at 1000 °C the solubility product [% Al] [% N] has the value 1.4×10^{-4} (equation (9.4.11)), while the solubility product [% V] [% N] according to equation (9.4.7) has the value 5.2×10^{-4}. When making non-ageing steel for deep drawing quality sheet, aluminium is sometimes replaced by vanadium because in this way better surfaces are obtained [1]).

In aluminium-killed steel the percentages of aluminium and nitrogen are usually such that the AlN present dissolves in the metal between 1000° and 1200 °C [2]). Heat-treatments must therefore be kept below this temperature region to prevent excessive growth of the γ crystals.

Aluminium nitride not only plays an important part in the control of austenite grain growth in aluminium-killed steels but also in that of the ferrite grain structure of aluminium-killed, low carbon steel sheets used for severe deep drawing and other difficult forming operations. The precipitation of AlN from the supersaturated solution, which takes place during the final heat-treatments before or during the recrystallization of the steel, has a strong influence on the shape and size of the crystals and hence on the mechanical properties of the material. Here again it is not only a question of the total quantity of precipitating AlN. To obtain the elongated grain shape required for deep drawing quality, the precipitate particles must have a suitable shape, size and distribution. For further technical particulars we refer to the extensive literature on this subject. We mention here only two representative articles [3,4]), the latter of which [4]) contains a comprehensive list of references.

THE ROLE OF Si_3N_4 IN MAKING GRAIN-ORIENTED SILICON STEEL SHEET

Until comparatively recently virtually all the silicon iron sheet for transformers (see Section 9.5) was obtained by hot-rolling. The directions of easy magnetization of the separate crystals in the sheet are then almost randomly

[1]) S. Epstein, H. J. Cutler and J. W. Frame, J. Metals 2, 830 (1950)
[2]) P. Werthebach and H. Hoff, Stahl Eisen 78, 736 (1958)
[3]) R. L. Solter and C. W. Beattie, Trans. AIME 191, 721 (1951)
[4]) H. Borchers, Z. Q. Kim and H. Hoff, Arch. Eisenhüttenwes. 35, 57 and 567 (1964)

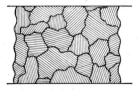

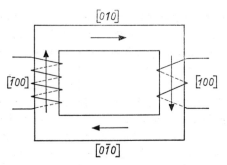

Fig. 124. The figure shows schematically that in hot-rolled silicon steel sheet the crystals show no preferential orientation with respect to the plane and direction of rolling.

Fig. 125. Symbolic representation of a transformer, the core of which is built up of single crystal sheets in such a way that the magnetic flux can everywhere follow a direction of easy magnetization.

distributed over the various directions in space (Fig. 124), so that the hysteresis losses are relatively large. In principle it would be most desirable to make the cores of large laminar single crystals (bounded by cube faces) in such a way that the magnetic flux always follows a direction of easy magnetization (Fig. 125).

Technically it is impossible to make single crystal sheet in large quantities. But it has been found possible to make on a large scale (in quantities of thousands of tons each month) polycrystalline silicon iron sheet, in which all the crystals have nearly the same orientation (Fig. 126). This orientation is such that the crystals lie with a (110) plane approximately parallel to the surface of the sheet and with a direction of easy magnetization, [001] direction, approximately parallel to the direction of rolling (Fig. 127). This texture is often referred to as Goss texture in honour of its inventor Goss, but also as (110)[001] or cube-on-edge texture [1-3]. The latter name is illustrated by Fig. 128, in which the orientation under discussion is demonstrated with the help of a number of cubes, which symbolize the unit cells. The Goss texture can only be obtained by cold-rolling silicon iron from a certain thickness and by subjecting it to certain heat-treatments.

For many years the way in which the crystal orientation in the Goss sheet is brought about was not understood and the production of the sheet did not always give the desired results. Our own experiments showed that it is

[1] N. P. Goss, Trans. Amer. Soc. Met. **23**, 511 (1935)

[2] R. M. Bozorth, Trans. Amer. Soc. Met. **23**, 1107 (1935)

[3] C. G. Dunn, *Cold Working of Metals*, Amer. Soc. for Metals, Cleveland 1949, p. 113-130.

Fig. 126. There is large-scale production of cold-rolled 3% silicon iron sheet in which all the crystals have about the same orientation, viz. the orientation (110) [001] (cf. Fig. 127).

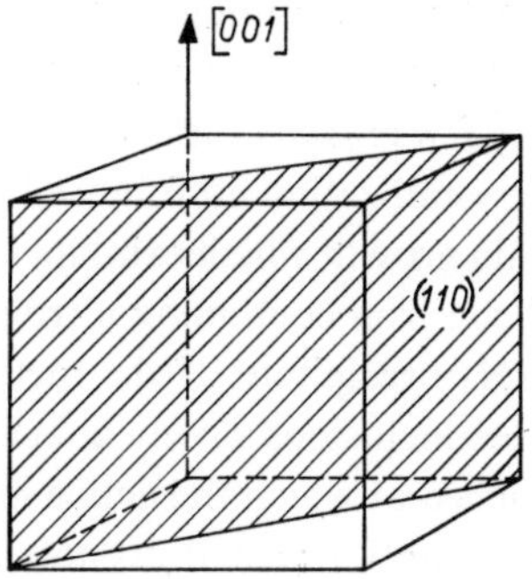

Fig. 127. In silicon iron sheet having Goss texture the crystals are orientated in such a way that they lie with a (110) plane approximately parallel to the rolling plane and with a [001] axis (cube axis) approximately parallel to the direction of rolling (cf. Fig. 128).

impossible to obtain the Goss texture in silicon iron sheet made from pure iron and transistor-quality silicon. From this we concluded that the presence of impurities in the material is of essential importance [1]. In further experiments we added measured quantities, in each case of one element, to pure silicon iron alloys. It was found that the desired texture is readily obtained by introducing nitrogen (in quantities of a few hundredths of a percent) and heat-treating the metal before cold-rolling in such a way that it contains a finely divided precipitate of Si_3N_4, which is mainly present at the grain boundaries [2]. Fig. 129 shows an electron microscopic photograph of 3% silicon iron, in which a precipitate of this type is present. (See Plate 7, opposite p. 278).

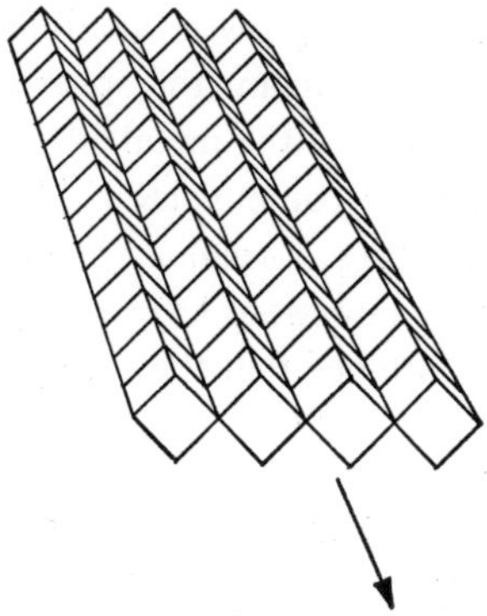

Fig. 128. Illustration by means of cubes of the crystal orientation in magnetic steel with Goss texture (cube-on-edge texture). The arrow indicates the direction of rolling.

[1] J. D. FAST, Philips Res. Rep. **11**, 490 (1956)
[2] See also: J. D. FAST and J. J. DE JONG, I. Physique Radium **20**, 371 (1959)

After cold-rolling and after primary recrystallization at 600° to 800 °C, both the pure silicon iron sheet and the sheet containing Si_3N_4 contain only very few crystals with the orientation (110)[001]. In the pure alloy the primary recrystallization is followed at high temperature (e.g. 900 °C) by normal grain growth which exhibits no preference for a particular orientation. In the silicon iron containing nitrogen the normal grain growth is inhibited by the Si_3N_4 precipitate. If the metal is heated in the temperature range in which these inclusions slowly coagulate and go into solution, then a point will be reached at which the active inclusions are so reduced in number that the few favourably oriented crystals can begin to grow but not, as yet, the others. In other words secondary recrystallization (exaggerated grain growth) occurs, by means of which a few primary crystals with orientation (110)[001] grow, at the cost of the other crystals, to many times the sheet thickness. The driving force for growth of these grains is the low gas-metal interfacial energy of the (110) surfaces in an atmosphere of pure hydrogen [1,2]. This surface energy is less than that of any (hkl) plane different from (110).

The selective grain growth under discussion occurs in a particular temperature range. If the nitrogen-bearing silicon iron is heated before or after the primary recrystallization immediately at too high a temperature (e.g. 1250 °C), then the active inclusions go into solution very rapidly, so that one obtains mainly normal grain growth and, as a consequence, a poor texture. Therefore one must first heat at a lower temperature (900°-1000 °C). As soon as the (110)[001] crystals have a sufficient start in size on the other crystals, the temperature can be raised to accelerate further growth.

The Si_3N_4 particles, which are so useful for producing the Goss texture, are unfavourable for the final magnetic characteristics of the material. They must therefore be removed from the silicon iron after they have accomplished their grain-growth function. This takes place automatically in the final heat-treatment in an atmosphere of pure hydrogen, since virtually all the nitrogen then leaves the metal [3,4].

In the commercial 3 % Si-Fe alloys, MnS is the most important impurity inhibiting normal grain growth after primary recrystallization [1]. Various other inclusions [5,6] can also perform this task. Here again it is of primary importance that they are present in the metal in the desired size and distri-

[1] J. E. MAY and D. TURNBULL, Trans. AIME 212, 769 (1958)
[2] J. L. WALTER and C. G. DUNN, Trans. AIME 215, 465 (1959) and 218, 1033 (1960)
[3] J. D. FAST, Philips Res. Rep. 11, 490 (1956)
[4] J. D. FAST and H. A. C. M. BRUNING, Z. Elektrochem. 63, 765 (1959)
[5] H. C. FIEDLER, Trans. AIME 221, 1201 (1961)
[6] M. J. MARKUSZEWICZ, J. Iron Steel Inst. 200, 223 (1962)

bution. For inclusions of MnS this has been shown most convincingly by Fiedler [1]). He experimented with a 3.3% silicon iron alloy which contained a little less than 0.1% MnS. This compound could be completely dissolved by heating the metal at 1325 °C. The most effective degree of dispersion could then be obtained in two ways: (a) by correct choice of the cooling rate, (b) by drastic quenching followed by precipitation heating at a lower temperature (1000 °C).

An advantage of Si_3N_4 over other grain-growth inhibitors is that after the final heat-treatment no impurities remain behind in the metal, so that the magnetic properties are particularly good.

The characteristics of silicon iron sheet with Goss texture deviate in some respects very little from those of single crystals. A disadvantage, however, is that the unfavourable [110] directions of the crystals lie in a direction perpendicular to the direction of rolling (cf. Fig. 127). One can therefore only take full advantage of this material if the magnetic flux is everywhere parallel to the direction of rolling, i.e. if it is used in the form of ring cores wound from sheet and not, for example, in the form of E sheets.

For many applications it would be very desirable to have available silicon iron sheet with cube texture [2-4]), i.e. sheet in which the crystals are so oriented that not only the direction of rolling, but also the direction perpendicular to it is a direction of easy magnetization (Fig. 130). In Germany and the U.S.A. they have already succeeded in making this material with (100)[001] texture on a small scale. In the production, the interaction between the metal surface and the surrounding gas atmosphere plays a very important part. If the oxygen activity of the gas exceeds a particular value, then it is no longer the (110) planes but the (100) planes which have a smaller

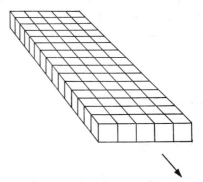

Fig. 130. Illustration of the cube texture by means of a number of cubes. The arrow indicates the direction of rolling.

[1]) H. C. Fiedler, Trans. AIME **230**, 95 (1964)
[2]) F. Assmus et al., Z. Metallkde **48**, 341 and 344 (1957)
[3]) J. L. Walter et al., J. appl. Phys. **29**, 363 (1958)
[4]) G. Wiener et al., J. appl. Phys. **29**, 366 (1958)

surface energy than all other (*hkl*) planes, in other words, the driving force for growth of (100) grains is then greater than that of (110) grains.

A convincing demonstration of the above is given by experiments of Walter and Dunn [1]) on the migration of (100)/(110) boundaries, i.e. boundaries between two grains, one of which has a (100) plane and the other a (110) plane parallel to the surface of the 3% silicon iron sheet. At 1200 °C the (100)/(110) boundaries advance into (100) grains in a good vacuum, then reverse their direction and migrate into (110) grains in an atmosphere of impure argon. The direction of migration reverses once again with (110) grains growing at the expence of (100) grains in a second vacuum anneal. These results are explained by the authors in terms of a change in concentration of oxygen atoms at the gas-metal interface during the anneals. The addition of oxygen atoms to the surface during the anneals in impure argon results in a decrease of the specific surface energy of the (100) oriented grains to a value lower than that of the (110) oriented grains. In a good vacuum or in pure hydrogen, however, the oxygen concentration at the surface is lowered to the point where the surface energy of the (110) grains has the lowest value.

The development of the cube texture in 3% silicon iron is, however, much more difficult and complicated than would be supposed from the foregoing. Control of the gas atmosphere in the final heat-treatment is a necessary, but not sufficient condition for success. Up to now it has not been found possible to produce silicon iron sheet with cube texture economically in large quantities. The main difficulty seems to be getting the alloy into such a condition that, after primary recrystallization of the sheet, there is a sufficient number of crystals present with the required (100)[001] orientation. According to patents of the General Electric Company (U.S.A.) this aim can be achieved by starting with ingots having favourably oriented columnar crystals obtained by controlled directional solidification.

9.7. Internal oxidation, nitriding and hydriding

INTRODUCTION

A finely dispersed second phase can be obtained in a number of alloys by allowing oxygen, nitrogen or hydrogen to diffuse into the metal from outside.

---

[1]) J. L. WALTER and C. G. DUNN, Acta Met. **8**, 497 (1960)

A condition for success is that the alloy contains a relatively small percentage of a substitutionally dissolved component which has a much greater affinity for oxygen, nitrogen or hydrogen than the base metal. A second condition which must be fulfilled is that the gas diffuses much more rapidly inwards than its reaction partner outwards. The conditions can then be chosen in such a way that the two react selectively with each other to form a finely divided and stable oxide, nitride or hydride. One then speaks of internal oxidation, internal nitrogenation (or nitriding) and internal hydrogenation (or hydriding). The advantages of the presence of these finely dispersed phases (hardening, inhibition of grain growth, etc.) have already been discussed in the previous section. The formation, there discussed, of a Si_3N_4 precipitate in silicon iron is an example of internal nitrogenation. The precipitate is obtained by heating silicon iron strip in a gas stream consisting of a mixture of hydrogen and ammonia, the ratio of the two being chosen such that no iron nitride can form. During heating nitrogen diffuses into the metal (cf. Sections 7.9 and 9.5) and at sufficiently high temperature reacts selectively with silicon, forming Si_3N_4.

The process of internal nitrogenation or nitriding is applied commercially on a large scale to harden certain ferrous alloys. The aim is not then to harden right through to the core of the material, but to harden a comparatively thin surface layer or case. This is known as case-hardening.

Internal hydrogenation or hydriding is used to harden magnesium, to which a small percentage of zirconium has been added. Internal oxidation has, as yet, found only a few applications, but can be used to harden certain silver and copper alloys.

NITRIDING

In contrast to the case-hardening process in which carbon is introduced into the surface layer of steel and which is known as carburizing [1]), hardening by nitriding does not come about via quenching and forming of martensite. The principal cause of hardening is here the above-mentioned formation of an extremely finely divided nitride. While carburizing of ferritic steels requires heating into the austenite region, i.e. to a temperature above the Ac_3 of the steel, for nitriding a temperature not much above 500 °C is desirable in order to prevent coalescence of the finely divided nitrides which form during this

---

[1]) See e.g. S. J. ROSENBERG and T. G. DIGGES, *Heat Treatment and Properties of Iron and Steel*, National Bureau of Standards Circular 495, Washington 1950.

heating. Any subsequent heat-treatment of the steel (quenching or tempering) is unnecessary.

Alloying elements with a great affinity for nitrogen, which may be considered in the development of steels for nitriding, include aluminium, chromium, titanium, zirconium and vanadium (cf. Table 1 in Chapter 3). The most used nitriding steels are the "nitralloys" which usually contain about 1% Al and 1% Cr as nitrogen binders. After nitriding these steels have very hard cases showing a great resistance to wear. Also, because the volume of the case increases as a result of nitride formation, surface layers have high compressive stresses which improve fatigue life [1]). Table 21 gives the compositions of a few nitralloys.

TABLE 21

COMPOSITION OF A FEW NITRALLOYS ACCORDING TO THE PERIODICAL "MATERIALS AND METHODS" (MARCH 1954, NUMBER 272)

Nitralloy type	135	135, modified	N	EZ	GR
Composition wt% (balance Fe)	C 0.30/0.40 Mn 0.40/0.70 Si 0.20/0.40 Cr 0.90/1.40 Al 0.85/1.20 Mo 0.15/0.25	C 0.38/0.45 Mn 0.40/0.70 Si 0.20/0.40 Cr 1.40/1.80 Al 0.85/1.20 Mo 0.30/0.45	C 0.20/0.27 Mn 0.40/0.70 Si 0.20/0.40 Cr 1.00/1.50 Al 0.85/1.20 Mo 0.20/0.30 Ni 3.25/3.75	C 0.30/0.40 Mn 0.50/1.10 Si 0.20/0.40 Cr 1.00/1.50 Al 0.85/1.20 Mo 0.15/0.25 Se 0.15/0.25	C 1.25/1.50 Mn 0.40/0.60 Si 1.25/1.50 Cr 0.20/0.40 Al 1.00/1.50 Mo 0.20/0.30

Because nitriding is carried out at a relatively low temperature, it is advantageous to quench and temper the steel before nitriding in order to get a strong, tough core beneath the hard, wear-resistant case. The strength of the core increases during nitriding if use is made of the nickel-bearing nitralloy N (Table 21) that age hardens at 500 °C [2]). This hardening process is due to phase separation. The effect is enhanced if the aluminium and nickel contents are increased to 2% Al + 5% Ni [3]).

Today, cams, gears, pinions, shafts, bushings, seals, cylinder barrels, clutches, piston rings and many other devices are hardened by nitriding. The aircraft industry makes extensive use of nitrided parts when lubrication is marginal. With requirements for wear and fatigue resistance at relatively high temperatures becoming more and more severe in many fields, it seems

[1]) See e.g. J. POMEY, Mém. Sci. Rev. Métallurgie 60, 215 (1963) and Härterei-Techn. Mitt. 18, 127 (1963)
[2])   NCH and V. O. HOMERBERG, Trans. Amer. Soc. Steel Treat. 20, 481 (1932)
[3]) W. S. MOUNCE and A. J. MILLER, Metal Progress, Febr. 1960, p. 91

certain that applications for nitriding will still increase in number in the near future [1]).

Hardening by nitriding can also be applied with success to many austenitic steels, e.g. to the well-known stainless steel containing 18% Cr and 8% Ni. In this case care must be taken that the oxide skin on the metal (to which it owes its resistance to corrosion) is removed before or during nitriding.

Nitriding can be carried out in either a liquid or a gaseous environment. Liquid nitriding takes place in a cyanide bath (NaCN or KCN). In the transfer of nitrogen to the metal, the presence of cyanate (NaCNO or KCNO) formed by oxidation plays an important part. Therefore various methods have been developed to deliberately increase the cyanate content of the bath ("activated nitriding", "tufftriding", etc.).

Gas nitriding has achieved wider use than liquid nitriding. This treatment takes place in an atmosphere of partly cracked ammonia. The process is sometimes carried out in gas at low pressure using a glow discharge. In that case a mixture of nitrogen and hydrogen may be used instead of ammonia. Nitriding by glow discharge offers various advantages. We shall mention here only that drilled holes can in this way be nitrided internally to a depth of more than ten times their diameter and that oxide films on stainless steels are automatically removed in the glow discharge by ion bombardment. For more thorough discussions of this process we refer to a number of recent articles [2-5]).

In all methods of nitriding difficulties may be experienced due to the formation on the surface of a "white layer", consisting mainly of iron nitride. It may cause brittleness and spalling in service. This skin can be removed by soaking the nitrided part in a hot solution of NaCN in water [6]) or by treating it in a glow discharge of pure hydrogen [3]). In some applications of nitriding the presence of a skin of iron nitride on the surface is considered quite desirable. In this case frequent use is even made of unalloyed steel [7,8]), so that strictly speaking one can no longer speak of "nitriding" in the sense in which we have used the word, viz. in the sense of "internal nitrogenation". After nitriding, unalloyed steel contains under the iron nitride skin a steel

[1]) W. LEEMING, Metal Progress, Febr. 1964, p. 86
[2]) H. HORNBERG, Härterei-Techn. Mitt. **17**, 82 (1962)
[3]) C. K. JONES and S. W. MARTIN, Metal Progress, Febr. 1964, p. 95
[4]) H. KNÜPPEL, K. BROTZMANN and F. EBERHARD, Stahl Eisen **78**, 1871 (1958)
[5]) T. M. NORÉN and L. KINDBOM, Stahl Eisen **78**, 1881 (1958)
[6]) D. A. DASHFIELD, Metal Progress, Febr. 1964, p. 88
[7]) E. MITCHELL and C. DAWES, Metal Treatment **31**, 3, 49, 88, 195 (1964)
[8]) J. A. PESCHAR, Metalen **18**, 98 (1963); in Dutch

layer in which a finely divided precipitate of iron nitride is present and under this again a steel layer containing nitrogen in supersaturated solution, the concentration of which decreases gradually from the outside to the inside.

HYDRIDING

It is known that the addition of about 0.6 wt % zirconium to magnesium inhibits the grain growth of this metal at high temperature to a considerable degree [1]). The associated high ductility of the binary alloy makes it attractive as canning material for fuel elements in nuclear reactors. It is already successfully used in thermal reactors in France. The mentioned content of 0.6 wt % is close to the solubility of Zr in liquid magnesium at temperatures between 655° and 800 °C [2,3]). Magnesium forms a peritectic system with zirconium, which implies that the solubility of Zr in solid Mg at the peritectic temperature (which is only a few degrees above the melting point of Mg) is greater than in liquid Mg. This maximum solubility is certainly greater than 1.0% and perhaps even much greater [4]). The solubility decreases rapidly with falling temperature and at 350 °C is already less than 0.1%. Unless cooling takes place extremely slowly, a 0.6% Zr-Mg alloy still retains much zirconium in supersaturated solution, since the diffusion coefficient of Zr in Mg is very small.

Zirconium performs its grain-growth inhibiting function in magnesium not only at temperatures at which a finely divided precipitate of metallic zirconium is present in the metal, but also at temperatures which are so high that the zirconium is present in solution. Ingots of magnesium with 0.6% Zr immediately after casting already show a much finer grain than ingots of magnesium without zirconium. Apparently this is a case analogous to the grain-refining action of aluminium in steel (see Section 9.6). Just as this effect is not produced by the aluminium as such but by an aluminium compound, so the grain-refining action of Zr in Mg must be ascribed to the presence of a finely divided zirconium compound, which is formed by reaction with an impurity.

Zirconium has a very strong affinity to oxygen, nitrogen and hydrogen and is therefore even used as "getter" in high-vacuum technique (see Section 9.8). One might therefore consider zirconium oxide, zirconium nitride or zirco-

[1]) F. SAUERWALD, Z. Metallkde 45, 257 (1954). In this article an extensive list of the literature on this subject will be found.
[2]) F. SAUERWALD, Z. anorg. allg. Chem. 255, 212 (1947)
[3]) J. H. SCHAUM and H. C. BURNETT, J. Research Nat. Bur. Standards 49, 155 (1952)
[4]) See also: E. F. EMLEY and P. DUNCUMB, J. Inst. Metals 90, 360 (1961/62)

nium hydride as the effective agent. Zirconium oxide can be eliminated from these compounds because magnesium has a greater affinity to oxygen than zirconium, so that ZrO_2 is reduced by Mg (see Table 1 in Chapter 3). On the other hand, zirconium has a stronger affinity for nitrogen and hydrogen than magnesium. In principle both zirconium nitride and zirconium hydride could function as grain-growth inhibitors. In practice the action of the hydride is of primary interest. This can be concluded from the facts that (a) Mg-Zr alloys during their manufacture undoubtedly react with water vapour from the gas atmosphere and furnace walls as given by the equation $Mg + H_2O \rightarrow$ $\rightarrow MgO + H_2$ [1]) and (b) intentional addition of hydrogen to the alloys strengthens the inhibition of grain growth (see below).

If one allows hydrogen to react with a liquid 0.6% Zr-Mg alloy, then nearly all the zirconium will precipitate in the form of a hydride which sinks to the bottom of the crucible [2,3]). If the zirconium-bearing magnesium is heated in the solid state in hydrogen, then a very finely divided hydride precipitate forms. As a result, not only the inhibition of grain growth, but also the mechanical resistance (hardness, tensile strength, yield point, resistance to creep) of the material increase. At the same time the electrical resistivity of the alloy decreases to nearly the value for pure magnesium. As a function of time the mechanical resistance passes through a maximum, which shows that hydride precipitation, as would be expected, is followed by coalescence of the precipitate. Taschow and Sauerwald [4]) measured for 0.6% Zr-Mg alloys after casting an average Brinell hardness of 41.2 kg/mm^2 and after 11 hours heating in hydrogen at 475 °C a value of 48.8 kg/mm^2. Extension of the heat treatment from 11 hours to 96 hours caused the average hardness to fall again to 41.6 kg/mm^2. With an optical microscope giving thousandfold magnification no precipitate could be observed after 11 hours heating, but it was visible after 96 hours.

The inhibition of grain growth is much less sensitive to the size of the precipitate particles than the mechanical resistance. It is still very effective even after coalescence. If the heating at 475 °C is not carried out in hydrogen, but in pure argon or CO_2, a precipitate of metallic zirconium will form which inhibits the grain growth to a much smaller degree than the hydride precipitate. This can be partly explained by the fact that a large part of the zirco-

[1]) Cf. G. T. HIGGINS and B. W. PICKLES, J. Nucl. Materials **8**, 160 (1963)
[2]) F. SAUERWALD, Z. Metallkde **40**, 41 (1949)
[3]) J. HÉRENGUEL, J. BOGHEN and P. LELONG, Comptes rendus **245**, 2272 (1957)
[4]) H. J. TASCHOW and F. SAUERWALD, Z. Metallkde **52**, 135 (1961)

nium remains in solution at 475 °C, while precipitation is almost complete when heating in hydrogen.

Investigation with the electron microscope [1]) has shown that the zirconium hydride in Mg-Zr alloys precipitates in the form of platelets with hexagonal symmetry. According to Lelong et al. (l.c.) these are identical to the platelets which form in zirconium with a low hydrogen content during slow cooling from 550 °C (see Section 8.2). According to the equilibrium diagram zirconium-hydrogen (Fig. 65) this would mean that in zirconium-bearing magnesium a precipitate of the cubic δ hydride is formed by reaction with hydrogen. Electron diffraction patterns by Harris et al. [2]) show, however, that the tetragonal ε hydride forms (see Section 8.1 and also an article by Whitwham [3])). This is consistent with the fact that there is an excess of hydrogen present during hydriding. The formation of ε hydride has been confirmed by Day [4]). Where the supply of hydrogen is restricted it is possible that δ zirconium hydride will be produced.

Precipitation of the hydride occurs both in the interior of the grains and at the grain boundaries. As is frequently observed with precipitation in alloys, in this case too there is a preference for the boundaries. In the crystals the precipitate platelets lie parallel to the basal plane of the magnesium with edges parallel to the $<11\bar{2}0>$ directions. The platelets are larger when they are formed at higher temperatures. The average diameter is 5 to 10 times as large after heating in hydrogen for 20 hours at 600 °C as it is after similar heating at 450 °C.

The coalescence of the precipitate at constant temperature takes place only very slowly, which is in keeping with the minute solubility of zirconium hydride in magnesium and with the low value of the diffusion coefficient of zirconium in the metal. Also in accordance with this is the fact that during continued heating in argon or CO_2 virtually no loss of hydrogen occurs. During the service life as canning material in CO_2 cooled nuclear reactors, the hydrogen content of the 0.6% Zr-Mg alloys will even be more inclined to increase than to decrease. This finds its explanation in the fact that it is impossible to prevent altogether the presence of water vapour, so that hydriding can take place via the reaction $Mg + H_2O \rightarrow MgO + 2$ [H]. The shape, size and distribution of the precipitate cannot be controlled and the

[1]) P. Lelong, J. Dosdat, J. Boghen and J. Hérenguel, Comptes rendus **251**, 2698 (1960) and J. Nucl. Materials **3**, 222 (1961)
[2]) J. E. Harris, P. G. Partridge, W. T. Eeles and G. K. Rickards, J. Nucl. Materials **9**, 339 (1963)
[3]) D. Whitwham, Mém. Scient. Rev. Métall, **57**, 1 (1960)
[4]) J. H. Day, J. Nucl. Materials **11**, 249 (1964)

properties of the alloy may be unfavourably influenced by this unintentional hydriding. It has therefore been proposed to deliberately pre-hydride the material [1]), choosing a treatment which would result in a beneficial distribution of precipitates.

INTERNAL OXIDATION

The strengthening of metals by internal oxidation was first observed by Meijering and Druyvesteyn [2,3]) during the second world war. They dissolved metals with a strong affinity for oxygen (Be, Mg, Al, Si, Ti, etc) in small concentrations in metals with a much smaller affinity for oxygen (Ag, Cu, Ni). By allowing oxygen to diffuse into these alloys a fine stable dispersion of oxide particles could be obtained. As in the processes of nitriding and hydriding already discussed it is of essential importance that the gas diffuses inwards much faster than the solute element outwards.

We are dealing with the simplest case when silver is chosen as the solvent metal. Its affinity for oxygen is so small that silver oxide (Ag_2O) cannot form on the metal above 190 °C at an oxygen pressure of 1 atm (see Section 4.1). However, according to Table 15 in Section 8.7, relatively large quantities of oxygen can dissolve in silver at high temperatures, while its diffusion coefficient in the metal is large. If silver sheet of 1 mm thickness, in which 0.3 wt% magnesium is homogeneously dissolved, is heated at 800 °C for 2 hours in air, its hardness is about 170 Vickers, while it is only about 35 Vickers after equal heating in hydrogen or nitrogen.

During heating of the 0.3 % Mg-Ag alloy in air the dissolved magnesium is converted into MgO. The progress of the internal oxidation as a function of time can be seen from the occurrence of a "case" in the cross-section of the sheet, which increases in depth with time of oxidation. This case becomes visible after polishing and etching because it etches more rapidly than the untransformed interior. The boundary between case and interior is quite sharp (see Fig. 131 on plate 7, opposite p. 278).

In the first stage of the internal oxidation single molecules of MgO will form. These cannot migrate as such in the metal. Coalescence of the oxide can only take place through dissociation of the molecules and diffusion of

---

[1]) P. GREENFIELD, unpublished work cited by J. E. HARRIS et al., J. Nucl. Materials 9, 339 (1963)
[2]) J. L. MEIJERING, Report of a Conference on Strength of Solids, held at the H. H. Wills Phys. Lab., University of Bristol on 7-9 July, 1947, published by the Physical Society, London 1948 (p. 140)
[3]) J. L. MEIJERING and M. J. DRUYVESTEYN, Philips Res. Rep. 2, 81 and 260 (1947)

the constituent atoms. The less stable the oxide formed, the more frequently will dissociation of its molecules take place. Thus, if the magnesium is replaced by a solute metal with a weaker affinity for oxygen, greater coalescence of the oxide and smaller hardening must be expected. This is in agreement with the experiments. Of interest, for example, are the results of experiments (Meijering, l.c.) in which magnesium was replaced by an equal atom percentage of cadmium, an element which, according to Table 1, has a much smaller affinity for oxygen than magnesium (at 25 °C, $\Delta G^0(CdO) = -54$ kcal/mole and $\Delta G^0(MgO) = -136$ kcal/mole). The cadmium-bearing silver exhibits no hardening whatsoever after internal oxidation. At 1200 times magnification round oxide particles are visible, and the oxidation boundary is vague; but in the hardened 0.3% Mg-Ag alloy no particles are visible at the highest magnification, and the case boundary is as sharp as a normal grain boundary.

That the MgO particles are extremely small can also be concluded from the facts that both the lattice constant and the electrical resistivity of the silver, both of which had already been increased by the presence of the dissolved magnesium, increase still further due to the internal oxidation. Thus the greatest portion of the MgO is probably present in the form of single molecules or particles consisting of only a few molecules. One can almost speak of a (strongly supersaturated) *solution* of MgO in silver. Even with the help of an electron microscope it is impossible, after the internal oxidation described, to observe MgO particles in the silver [1]).

The replacement of silver as solvent metal by copper will facilitate the coalescence of the oxide formed by internal oxidation. The frequency of dissociation of the "dissolved" oxide molecules, e.g. the frequency of the reaction

$$[MgO] \rightarrow [Mg] + [O],$$

is not exclusively determined by the affinity of the solute metal (in this case Mg) for oxygen, but rather by the difference in the affinities of the solute metal and the solvent metal for oxygen. If both metals had the same affinity for oxygen, there would be no preferential oxidation at all. To achieve hardening of binary copper alloys by internal oxidation, more stringent requirements must be made for the solute metal with regard to its affinity for oxygen than in the case of silver alloys. For instance, it is instructive that silver with a small percentage of manganese can be hardened by internal oxidation, while copper with the same atomic percentage of manganese can

---

[1]) J. L. Meijering and G. Baas, unpublished work

not. The oxide particles formed are much larger in the latter case than in the former. Thus, while for the occurrence of internal oxidation it is only necessary for the affinity of the solute metal for oxygen to be greater than that of the solvent metal, for the occurrence of hardening by internal oxidation the difference in the affinities must be *large*.

The fact that copper has a greater affinity for oxygen than silver has another consequence than that discussed above. During oxidation in air a layer of oxide (chiefly Cu_2O) forms on the metal. The corresponding loss of metal can be reduced by heating the alloy in air only as long as is necessary for the formation of a Cu_2O layer sufficiently thick to supply the oxygen required for internal oxidation of the remaining metal. Heating is then continued in argon or another inert gas.

As an example of the hardening of a copper alloy by internal oxidation it can be mentioned that the hardness of copper containing 0.2 wt% beryllium increases in this way from 40 to 165 Vickers. The internal oxidation of copper alloys was observed by Smith [1] as early as 1930 and studied in detail metallographically by Fröhlich [2] and Rhines [3]). The strengthening which occurs in some of these alloys was discovered by Meijering and Druyvesteyn (l.c.) after they had first predicted and demonstrated the strengthening of silver alloys by internal oxidation.

The alloys 0.3% Mg-Ag and 0.2% Be-Cu, after internal oxidation are not only very hard, but also show marked intergranular brittleness, which is probably caused by the presence of MgO or BeO at the grain boundaries. Addition of 0.1% Ni to the 0.3% Mg-Ag alloy gives (as in the case of magnesium-free silver) a much finer grain and, at the same time, reduces considerably the grain boundary brittleness after internal oxidation. Since nickel has only a minute solubility in silver, the effect probably depends for the greater part on a high concentration of nickel at the grain boundaries which makes it more difficult for MgO to form there [4]). The internally oxidized alloy Ag + 0.3% Mg + 0.1% Ni has found applications as a material for electrical contacts.

Also nickel and iron, if they contain a small percentage of e.g. aluminium, can be internally oxidized. Nickel and, in particular, iron have, however, such a relatively large affinity for oxygen that *hardening* by internal oxidation does not occur here under normal conditions. Only by starting from powders

[1] C. S. SMITH, Min. and Met. **11**, 213 (1930), **13**, 481 (1932)
[2] K. W. FRÖHLICH, Z. Metallkde **28**, 368 (1936)
[3] F. N. RHINES, Trans. AIME **137**, 246 (1940)
[4] J. A. SLADE and J. W. MARTIN, Metallurgia **64**, 23 (1961)

can interesting hardening phenomena be obtained in this case (see end of this section). Fig. 132 shows the internally oxidized zone in an alloy of the composition 69% Fe + 30% Ni + 1% Al [1]). It was obtained by heating the alloy for $\frac{1}{4}$ hour at 1200 °C in 80% argon + 20% oxygen and subsequently for $5\frac{3}{4}$ hours in pure argon. The choice of composition of the alloy is based on the fact that after cooling to room temperature it still has the same structure (austenitic) as during the oxidation heating at 1200 °C. Complications due to crystallographic transitions are thus avoided. The oxide particles which can be seen in Fig. 132 are too coarse to cause hardening. It is possible that the coarsest particles near the outer surface consist of $(Fe,Ni)Al_2O_4$ and the less coarse ones near the inner boundary of the case of Al_2O_3. A binary alloy 99% Fe + 1% Al after internal oxidation shows only the latter state of affairs, i.e. that of the inside of the case in Fig. 132.

Fig. 133 shows the internally oxidized zone of an alloy of the composition 94.5% Fe + 5% Sn + 0.5% Al. It was obtained in the same way as the zone in Fig. 132. The choice of composition of the tin-bearing alloy is based on the fact that its structure in the entire temperature range between room temperature and 1200 °C is ferritic. Figs. 132 and 133 show clearly that the diffusion of oxygen in austenitic and ferritic steel shows no preference for the grain boundaries: the boundary of the internally oxidized rim is parallel to the metal surface and shows no bulges where it cuts grain boundaries. The same is true for Ag, Cu and Ni alloys under the same conditions of a relatively small concentration of the solute element and not too low an oxidation temperature.

It also follows from the above that internal oxidation is an exceptionally sensitive method of investigating whether oxygen dissolves in a metal. This turns out to be the case in iron, although the solubility in α and γ iron is so small that it cannot be demonstrated with certainty in any other way, not even by means of the very sensitive method of internal friction (see also Sections 4.5 and 7.13). Meijering [2]) even succeeded by means of the method of internal "oxidation" in showing that fluorine can dissolve and diffuse in silver.

It may seem surprising that iron with about 1% Al can be hardened by internal *nitriding* (see above), but not by internal *oxidation*. If one only takes into consideration the free enthalpies of formation of AlN and Fe_4N and those of $\frac{1}{2}Al_2O_3$ and $\frac{3}{2}FeO$, then one would be more

[1]) J. L. MEIJERING, Acta Met. **3**, 157 (1955)
[2]) J. L. MEIJERING, Mém. Scient. Rev. Métall. **54**, 520 (1957)

PLATE 7

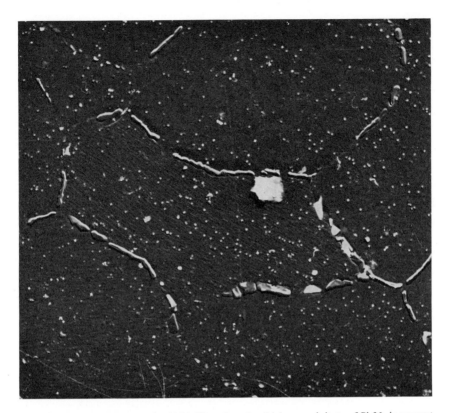

Fig. 129. Electron micrograph of 3 % silicon iron in which a precipitate of Si₃N₄ is present along the grain boundaries. The micrograph corresponds to an area of 17×19 microns.

Fig. 131. Cross-section of silver strip (1 mm thick) containing 0.3 % Mg, after heating for 50 minutes at 800 °C in air and etching in $NH_4OH + H_2O_2$. Magn. $10 \times$. In the "case" (internally oxidized zone) the magnesium has been converted into MgO, but not yet in the interior. The boundaries of the separate crystals can be seen in the photograph. The MgO is too finely divided to be made visible (Meijering).

PLATE 8

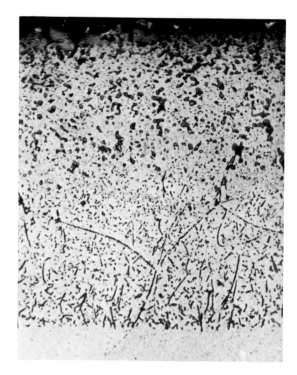

Fig. 132. Internally oxidized zone in an alloy of composition 69% Fe + 30% Ni + 1% Al after heating for ¼ hour at 1200 °C in 80% argon + 20% oxygen and 5¾ hour at 1200 °C in pure argon. Etched with $(NH_4)_2S_2O_8$. Magnification 275 × (Meijering).

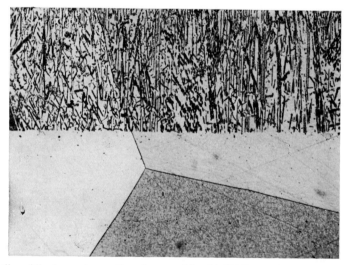

Fig. 133. Internally oxidized zone in an alloy of composition 94.5% Fe + 5% Sn + 0.5% Al. Same heat-treatment as in Fig. 132. Etched with nital. Magnification 200 × (Meijering).

inclined to expect the reverse. Nevertheless, one must also take into account that AlN and Al_2O_3 are formed in the first instance in a state of "solution" in the iron. Al_2O_3 molecules will certainly produce so much greater distortion of the iron lattice than AlN molecules, that the heat of solution of $\frac{1}{3} Al_2O_3$ in iron will have a much larger positive value than that of AlN. The result is that the dissolved Al_2O_3 molecules have a greater tendency to dissociate than AlN molecules. The greater dissociation leads to more rapid coalescence and reduced hardening. Furthermore, according to Meijering [1]), one must take into account the fact that the reaction [Al] + [N] → [AlN] in iron requires a considerable activation energy (perhaps about 40 kcal), as is shown by the vague boundary between case and interior of the metal. In the other direction of the reaction this will retard dissociation and conglomeration.

If the concentration of the solute element to be oxidized exceeds a certain value (which depends, amongst other things, on the solubility of oxygen in the solvent metal), then selective oxidation of this element does not produce separate particles, but oxide films in the interior of the metal. At even greater concentrations only an oxide layer forms at the outer surface [2,3]).

For recent studies of the mechanical properties of internally oxidized silver and copper alloys we refer to a number of articles [4-8]). Spectacular hardening effects have been obtained by Grant and his co-workers [9,10]) who produced internally oxidized powders of copper and nickel alloys and compacted them by a process of hot extrusion. The strength of these alloys at high temperatures is extremely large.

9.8. Interaction of Th₂Al and related getters with hydrogen

INTRODUCTION

In high-vacuum technique frequent use is made of metals and alloys whose task it is to bind gases. Particularly in the manufacture of electron tubes these "getters" play an important role.

[1]) J. L. MEIJERING, Bristol Conference, 1947 (l.c.)
[2]) C. WAGNER, Z. Elektrochem. 63, 772 (1959)
[3]) F. MAAK, Z. Metallkde 52, 538 and 545 (1961)
[4]) J. W. MARTIN and G. C. SMITH, J. Inst. Metals 83, 153 and 417 (1954/55)
[5]) E. GREGORY and G. C. SMITH, J. Inst. Metals 85, 81 (1956/57)
[6]) M. H. LEWIS and J. W. MARTIN, Acta Met. 11, 1207 (1963)
[7]) M. J. MARCINKOWSKI and D. F. WRIEDT, J. Electrochem. Soc. 111, 92 (1964)
[8]) H. SPENGLER, Metall 18, 727 (1964)
[9]) O. PRESTON and N. J. GRANT, Trans. AIME 221, 164 (1961)
[10]) L. J. BONIS and N. J. GRANT, Trans. AIME 224, 308 (1962)

In the early years of the radio industry mainly magnesium was used as getter. After the tubes had been pumped to a certain pressure, the magnesium was evaporated causing the gases still present to be removed much more rapidly and to a lower final pressure than was possible by means of the high-vacuum pumps. Nowadays, interest is centred almost exclusively on getters which not only shorten the pumping time of a tube, but which are also able to bind the gases which, after sealing off, are liberated from the components of the tube during its service life. The requirement is thus made that the getter shall act during the life of the tube as a high-vacuum pump continually connected to it. Magnesium does not fulfill this requirement, mainly because it does not take up hydrogen.

The gases which escape from the glass, ceramic and metal parts of an electron tube are chiefly hydrogen, water vapour, carbon monoxide, carbon dioxide and hydrocarbons [1]. To clean up these gases use is now made on a large scale of barium which is deposited on the glass walls by evaporation. Getters like barium are called evaporating or "flashed" getters as contrasted to the non-evaporating or "non-flashed" getters.

Examples of non-evaporating getters are zirconium and titanium. Both metals can be evaporated, but normally they are used unevaporated in the form of sheet or wire or more often as a powder fixed at appropriate parts of the tube by means of electrophoresis or with the help of a binder.

SORPTION CAPACITY, SORPTION RATE AND RESIDUAL PRESSURE

The quality of a getter is determined by three factors:
(1) The sorption *capacity* for the gases liberated in vacuum tubes,
(2) The *rate* at which these gases are taken up,
(3) The *residual pressures* of the gases with which the getter is in equilibrium at its operating temperature.

These three factors are interdependent in the sense that the sorption rate must be such that very low residual pressures of the gases must be reached in relatively short time. These residual pressures depend, like the sorption rates, on the quantities of gas already taken up. This means that the sorption capacity depends on the requirements which are made for the vacuum.

The behaviour of the getter towards hydrogen is of primary importance. The hydrides are always the least stable compounds that getters form with gases. As a consequence they have the highest dissociation pressures. Particularly at temperatures where the sorption rates for the other gases

---

[1] A. KLOPFER, S. GARBE and W. SCHMIDT, Vacuum **10**, 7 (1960)

become sufficiently large, the hydrogen equilibrium pressure may be too high. In some applications, for example where oxide cathodes are concerned, a certain hydrogen pressure may be desirable in the tube. One then uses a selective getter, one which binds oxygen, nitrogen and carbon and whose temperature corresponds to a small equilibrium pressure of hydrogen.

ACTIVATION OF GETTERS

Not every material that is sufficiently reactive with respect to the gases to be cleaned up can be used as a getter. It is essential that one must be able to determine the moment at which the getter begins to work. If this is not possible it "corrodes" in such a way during the manufacturing process (before sealing off) that it is unfit at the time when it is required to perform its work. The putting into action of the getter is called activating. Before activation the getter is prevented from doing its work either by mechanical or chemical means. Barium, for example, in non-activated state is present in an electron tube either in a small piece of nickel tube with both ends squeezed or in the form of the non-reactive intermetallic compound $BaAl_4$ in a tiny iron "boat". The former case is one of mechanical protection, the latter of chemical protection. In both cases activation occurs by heating. Barium either evaporates through the slits in the ends of the nickel tube or is liberated by the reaction of $BaAl_4$ with iron, and is deposited on the glass wall as an active layer.

In the case of non-evaporating getters, protective surface layers prevent them from doing their getter work too soon. On titanium and zirconium these are the oxide films which form spontaneously in the air. By heating these metals at, say, 1000 °C the oxide dissolves in the metal (see Section 8.5). The metal is then in an activated state and even after cooling can function as getter.

MODERN NON-EVAPORATING GETTERS

In some small types of electron tubes evaporating getters cannot be employed because, among other reasons, the metal can condense in places where its presence is undesirable. One is then forced to use non-evaporating getters. Like other getters they must satisfy the condition that their affinity for hydrogen, oxygen, nitrogen and carbon is great, i.e. that their hydrides, oxides, nitrides and carbides (or the solutions of H, O, N and C in the metal) are very stable.

As an example of a well investigated non-evaporating getter, which in many respects works better than titanium or zirconium, we shall discuss the Ceto-getter [1]). For a description of another modern type of non-evaporating getter, the porous-type getter, the reader is referred to the literature [2]).

Ceto is a brittle alloy containing principally the metals thorium, cerium and aluminium. After its preparation by sintering it is crushed under a protecting liquid, mixed with a binder to a paste and subsequently painted on the proper parts of the valve. During a part of the manufacturing process the binder acts as a protecting layer.

THE TERNARY SYSTEM Th-Al-Ce

To obtain more insight into the functioning of the Ceto getter the binary systems Th-Al, Al-Ce and Ce-Th and also the ternary system Th-Al-Ce were thoroughly investigated by Van Vucht [3]) by X-ray, metallographical and thermoanalytical methods. Thorium and cerium form an uninterrupted series of fcc solid solutions. In the Th-Al and Al-Ce systems various intermetallic compounds occur, of which Th_2Al exhibits interesting getter properties. In this compound about 20% of the thorium atoms can be replaced in a random manner by cerium atoms without disturbing the homogeneity. Van Vucht (l.c.) was able to show that this homogeneous ternary phase with $CuAl_2$ structure is the principal component and at the same time the most active component of the Ceto getter. Fig. 134 shows the tentative diagram of the Th-Ce-Al system at 500 °C.

> The Ceto getter is rather more complicated than has been sketched above. In its preparation one does not use cerium, but "mixed metal" which is a mixture of about 80% cerium, 19% lanthanum and 1% other rare-earth metals. Van Vucht's experiments, however, show that a replacement of "mixed metal" by pure cerium yields a product that is essentially identical with Ceto. His investigations are therefore restricted mainly to the ternary system, but all the conclusions drawn from the work on "pure Ceto" have been checked by experiments with the factory product.

ABSORPTION OF HYDROGEN BY Th_2Al

The tetragonal unit cell of Th_2Al contains eight thorium atoms and four aluminium atoms. In addition to other interstices this cell contains sixteen

[1]) W. Espe, M. Knoll and M. P. Wilder, Electronics 23, 80 (1950)
[2]) N. Hansen, Vakuum-Technik 12, 167 (1963)
[3]) J. H. N. van Vucht, Philips Res. Rep. 16, 1 (1961)

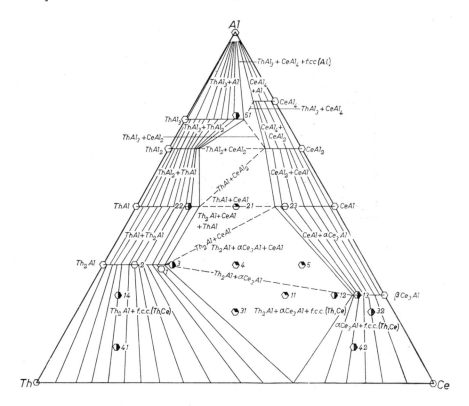

Fig. 134. Tentative diagram of the Th-Ce-Al system at 500 °C. The tie-lines shown in the figure are only schematic. The composition of Ceto corresponds to point 1 (Van Vucht).

equivalent interstices which are surrounded tetrahedrally by Th atoms and occur in pairs, as is shown in Fig. 135. It is seen from the figure that each couple of tetrahedra has one plane in common, viz. the plane which lies parallel to the base of the unit cell. The compound can absorb hydrogen up to about the composition $Th_8Al_4H_{16}$. From neutron diffraction diagrams of Th_2Al containing deuterium and proton magnetic resonance measurements with Th_2Al specimens containing hydrogen Van Vucht [1]) was able to conclude that in $Th_8Al_4H_{16}$ all sixteen of the above-mentioned interstices contain an H atom, while the four remaining interstices are empty.

Van Vucht [2]) determined the H_2 equilibrium pressures as a function of the

[1]) J. H. N. VAN VUCHT, Philips Res. Reports **18**, 35 (1963)
[2]) J. H. N. VAN VUCHT, Philips Res. Reports **18**, 1 (1963)

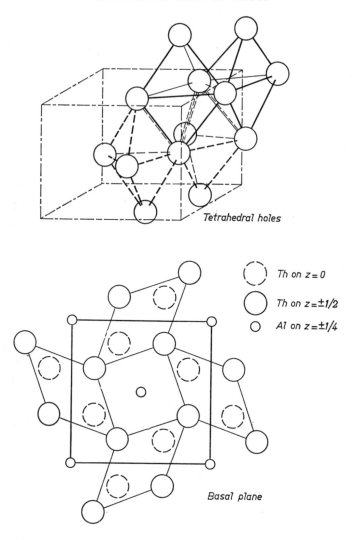

Fig. 135. Structure of Th₂Al. The arrangement of tetrahedral holes as twin interstices is clearly visible (Van Vucht).

hydrogen content of Th₂Al at a number of temperatures. Fig. 136 gives the results of these measurements. The isotherms do not show any kind of plateau, as might be expected if at these temperatures a second solid phase were present. Strictly speaking, we are dealing with a ternary system (Th-Al-H), so that the absence of a plateau does not necessarily imply that no new phase is formed. However, the temperatures of the isotherms are low compared

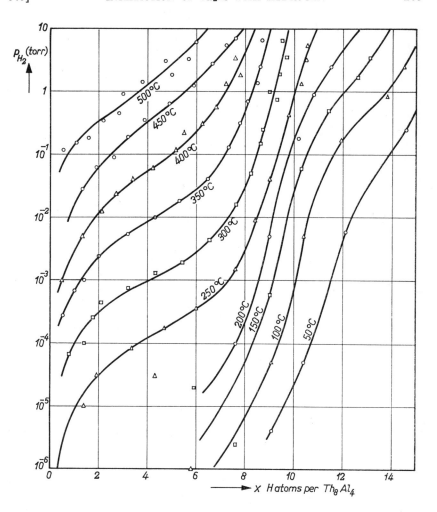

Fig. 136. Isotherms in the Th₂Al-hydrogen system. The logarithm of the hydrogen pressure in mm of mercury (torr) has been plotted against the H/Th₈Al₄ ratio (Van Vucht).

with the melting point of Th₂Al (1307 °C) and, as a consequence, aluminium and thorium atoms are unable to migrate appreciably in the time available for the experiments. Phase separation into a mixture of ThH₂ and aluminium, which might be expected on thermodynamic grounds, cannot take place for kinetic reasons. It therefore is admissible to consider the system as quasi-binary and, according to Gibbs' phase rule, the absence of a region of constant pressure in the isotherms may be interpreted to mean that the solid phase remains homogeneous at all hydrogen contents.

Standard values of the relative partial molar enthalpy and relative partial molar entropy of hydrogen in the metallic phase can be deduced from the pressure measurements in a manner similar to that described in Section 8.3 for the niobium-hydrogen system. It will be remembered that these are the changes in enthalpy and entropy which take place when 0.5 mole H_2 at 1 atm is isothermally dissolved in an infinitely large quantity of the solid solution in question. For 250 °C these values are to be found in Table 22, which may be compared with Table 14 in Section 8.3.

TABLE 22

VALUES OF THE RELATIVE PARTIAL MOLAR ENTHALPY AND RELATIVE PARTIAL MOLAR ENTROPY OF HYDROGEN IN Th_2Al AT 523 °K ACCORDING TO VAN VUCHT.

H/Th_8Al_4	$\Delta \bar{h}_H$ cal per gr.-atom H	$\Delta \bar{s}_H$ cal/degr. per gr.-atom H
1	-13900	-8.5
2	-14100	-10.0
3	-14500	-11.4
4	-15500	-13.5
5	-16300	-15.2
6	-17100	-17.0
7	-17300	-18.1
8	-17600	-19.3

The partial heats of solution (relative partial molar enthalpies, $\Delta \bar{h}$) of hydrogen in Th_2Al are seen from Table 22 to become more strongly negative with increasing hydrogen content in the concentration range from $Th_8Al_4H_1$ to $Th_8Al_4H_8$. Clearly, a dissolving hydrogen atom prefers the neighbourhood of other hydrogen atoms already present. This implies that they will tend to cluster at temperatures low enough for the entropy to become relatively unimportant. In other words: in the region of concentrations under consideration we have to expect a separation into two phases at lower temperatures. According to these expectations one of these phases will have a small concentration of hydrogen and the other will contain the majority of the hydrogen. In Chapter 8 (especially Section 8.3) this phenomenon has already been discussed for two metallic elements (Pd and Nb), in which hydrogen also dissolves exothermally.

PHASE SEPARATION AND ORDERING

The isotherms in Fig. 136 show, in the region of 225 °C, no less than three points of inflection. Two of these, lying at approximately $Th_8Al_4H_4$ and $Th_8Al_4H_{12}$, were regarded by Van Vucht as the vestiges of plateaus occurring at lower temperatures and corresponding to two-phase regions. This would imply that the composition $Th_8Al_4H_8$ should be considered as an intermediate hydride. X-ray diffraction showed that at room temperature and relatively small hydrogen concentrations separation into two phases does, in fact, occur, the composition of the phases being approximately $Th_8Al_4H_0$ and $Th_8Al_4H_5$. This is in satisfactory agreement with the expectations derived from Table 22. A similar region between $Th_8Al_4H_8$ and $Th_8Al_4H_{16}$ could not, however, be demonstrated in this way, even at a temperature as low as 83 °K [1]).

The changes in dimensions of the unit cell with increasing hydrogen content could be interpreted by Van Vucht by assuming that at the composition $Th_8Al_4H_8$ the hydrogen atoms are present in a more or less orderly manner in the interstices. From a theoretical analysis of the hydrogen-absorption isotherms [2]) it could be concluded that the hydrogen atoms have, in fact, a strong tendency to occupy each pair of interstices with only a single hydrogen atom. At the composition $Th_8Al_4H_8$ an intermediate state has been attained in which at low temperatures all the "double holes" contain one single proton.

EFFECT OF CERIUM CONTENT

Replacement of part of the thorium atoms in Th₂Al by cerium atoms increases the stability of the hydrogen solutions. This is demonstrated by equilibrium pressure measurements (similar to those described above) with specimens of the compositions $(0.875 \text{ Th, } 0.125 \text{ Ce})_8Al_4$ and $(0.75 \text{ Th, } 0.25 \text{ Ce})_8Al_4$. X-ray analysis shows that the first of these is fully homogeneous and of a structure identical with that of Th₂Al, whereas about 90% of any specimen of the second composition (which is very near to that of the getter Ceto) has the Th₂Al structure.

Figs. 137 and 138 give the results of the pressure measurements. The latter shows clearly the occurrence of plateaus. Judging from their appearance, the critical temperature, above which there is only one homogeneous phase, must be very high.

[1]) J. H. N. VAN VUCHT, Philips Res. Reports **18**, 21 (1963)
[2]) J. H. N. VAN VUCHT, Philips Res. Reports **18**, 53 (1963)

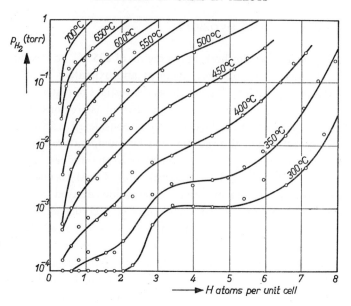

Fig. 137. Isotherms of hydrogen in equilibrium with (0.875 Th, 0.125 Ce)$_8$Al$_4$. In this figure and the following one the logarithm of the hydrogen pressure in mm of mercury (torr) has been plotted against the number of hydrogen atoms per unit cell (Van Vucht).

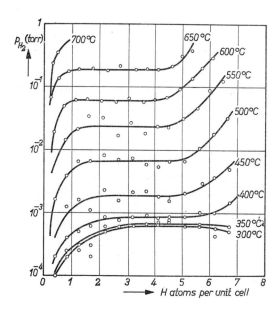

Fig. 138. Isotherms of hydrogen in equilibrium with (0.75 Th, 0.25 Ce)$_8$Al$_4$ (Van Vucht).

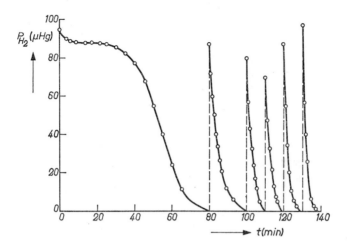

Fig. 139. Hydrogen sorption at room temperature by Th₂Al. After sorption of the first amount of hydrogen, new portions of gas were admitted successively, without changing the other conditions. It will be seen that each new portion is taken up more rapidly than the previous portion (Van Vucht).

The replacement of part of the thorium atoms in Th₂Al by Ce atoms results not only in an increase in the thermodynamic stability of the hydrogen-bearing alloys. Of greater importance is the fact that hydrogen is absorbed more rapidly. The explanation lies in the fact that the oxide skin which forms spontaneously on the surface of cerium-bearing alloys when water vapour and other gases are liberated, permits the passage of hydrogen more easily than the oxide which forms on pure Th₂Al. Once sufficient hydrogen has been absorbed, the accompanying expansion of the alloy causes cracks in the oxide skin. One therefore observes an autocatalytic acceleration of the hydrogen absorption. This is valid for both the cerium-bearing metal and for pure Th₂Al (Fig. 139). For further particulars on these kinetic questions we refer to an article by Van Vucht [1].

[1] J. H. N. VAN VUCHT, Philips Res. Reports **16**, 245 (1961)

INDEX